5G Networks & Cybersecurity

Published 2026 by River Publishers
River Publishers
Broagervej 10, 9260 Gistrup, Denmark
www.riverpublishers.com

Distributed exclusively by Routledge
605 Third Avenue, New York, NY 10017, USA
4 Park Square, Milton Park, Abingdon, Oxon OX14 4RN

5G Networks & Cybersecurity / Anand R. Prasad.

Routledge is an imprint of the Taylor & Francis Group, an informa business

ISBN 978-87-438-1088-9 (paperback)
ISBN 978-87-438-1091-9 (online)
ISBN 978-8-743-81090-2 (ebook master)
ISBN 978-87-438-1093-3 (ePub)
DOI: 10.1201/9788743810902

5G Networks & Cybersecurity

Editor

Anand R. Prasad
Deloitte Tohmatsu Cyber (DTCY),
Japan

River Publishers

Routledge
Taylor & Francis Group
NEW YORK AND LONDON

Contents

Preface

As 5G continues to reshape the global communication landscape, security has emerged as a main concern. With expanded connectivity, support for multiple services and devices, and the increasing dependence on virtualized and cloud-based infrastructure, the need to understand the complexities of 5G security is foremost. This book explores key elements of 5G security, from foundational principals to practical application related considerations and regulatory impacts.

- Chapter 1 sets the foundation by introducing the core principles of 5G and provides an overview of the evolving security landscape.
- Chapter 2 delves into various authentication mechanisms in 5G, describing how they are structured to meet the needs of various users and device types.
- Chapter 3 digs into how 5G security features are practically implemented, highlighting optional controls, operator decisions and the network security associated with them.

Aims and Objectives

This book aims to provide readers with a clear and structured understanding of 5G security architecture, its basic mechanism and the limitations that affect its real-world implementation. By combining technical insights with the regulatory context, this book is designed for professionals, researchers and policy makers seeking to understand how 5G security is designed, adapted and managed in today's complex digital infrastructure.

List of Authors

Anand R. Prasad, *Partner at Deloitte Tohmatsu Cyber (DTCY), Japan*

Sivabalan Arumugam, *Rakuten Mobile Inc, Japan*

Takahito Yoshizawa, *COSIC, KU Leuven, Belgium*

Sheeba Backia, *Lenovo, Motorola Mobility, Germany*

Mary Baskaran, *Lenovo, Motorola Mobility, Germany*

Xiaoting Huang, *China Mobile Research Institute, China*

Hans C. Rudolph, *Deloitte Tohmatsu Cyber, Japan*

List of Notations and Abbreviations

3GPP	3rd Generation Partnership Project
5G	5th generation
5G AKA	5G Authentication & Key Agreement
5G-RG	5G residential gateway
5GS	5G system
5G-GUTI	5G global unique temporary UE identifier
5G NSA	5G non-standalone
5G SA	5G standalone
AAA	Authentication, authorization, and accounting
AAA-P	Proxy AAA server
AAA-S	Serving AAA server
AAnF	AKMA anchor function
ABBA	Anti-bidding down between architectures
AF	Application function
AKA	Authentication and key agreement
AKID	AKMA key identifier
AKMA	Authentication and key management for applications
AMF	Access and mobility management function
AP	Access point
API	Application programming interface
ARPF	Authentication credential repository and processing function
AS	Access stratum
AUN3	Authenticable non-3GPP devices
AUSF	Authentication server function
AUTN	Authentication token
AV	Authentication vector
BBF	Broadband forum

BEST	Battery efficient security for very low throughput MTC devices
CK	Ciphering key
CRA	Cyber Resilience Act
DN	Data network
DNN	Data network name
DTLS	Datagram transport layer security
EAP	Extensible authentication protocol
EAP-AKA'	Extensible authentication protocol for AKA
EMSK	Extended master session key
ENISA	European Union Agency for Cybersecurity
ENSI	External network slice information
EUCC	EU Common Criteria Certification
EU5G	EU 5G Certification
FN	Fixed network
FN-RG	Fixed network residential gateway
GBA	Generic bootstrapping architecture
GPSI	Generic public subscription identifier
gNB	5G NodeB
HPLMN	Home PLMN
IaaS	Infrastructure as a service
IK	Integrity key
IMSI	International Mobile Subscription Identity
IPSec	Internet protocol security
IPUPS	Inter-PLMN user plane security
ME	Mobile equipment
MM	Mobility management
MTC	Machine type communication
N5CW	Non-5G-capable over WLAN
N3IWF	Non-3GPP inter-working function
NAI	Network access identifier
NAS	Non-access stratum
NDS	Network domain security
NESAS	Network equipment security assurance scheme
NF	Network function
NFV	Network function virtualization
ngKSI	Next-generation key set identifier
NGMN	Next generation mobile network
NPN	Non-public network
NSSAAF	NSSA application function
NSSAI	Network slice selection assistance information

NSWO	Non-seamless WLAN offload
NSWOF	Non-seamless WLAN offload function
OTA	Over-the-air
PaaS	Platform as a service
PLMN	Public land mobile network
PNI	Public network Integrated
PPP	Point-to-point protocol
PRINS	PRotocol for N32 interconnect security
RAN	Radio access network
RAND	Random number
RES	Response
REST	Representational state transfer
RG	Residential gateway
RID	Routing indicator
SBA	Serviced based architecture
SBI	Service-based interface
SDN	Software defined network
SEAF	Security anchor function
SEPP	Security edge protection proxy
SIDF	Subscription identifier de-concealing function
SMC	Security mode command/security mode complete
SMF	Session management function
SNN	Serving network name
SNPN	Standalone NPN
S-NSSAI	Single network slice selection assistance information
SoR	Steering of roaming
SQN	Sequence number
SUCI	Subscription concealed identifier
SUPI	Subscription permanent identifier
TNAN	Trusted non-3GPP access network
TNAP	Trusted non-3GPP access point
TNGF	Trusted non-3GPP gateway function
TR	Technical report
TS	Technical specification
TWAN	Trusted WLAN access network
TWAP	Trusted WLAN access point
TWIF	Trusted WLAN interworking function
UAS	Uncrewed aerial system
UAS-NF	Uncrewed aerial system network function
UDM	Unified data management
UDR	Unified data repository

UE	User equipment
UPF	User plane function
UPU	UE parameter update
USIM	Universal mobile identity module
USS	UAS service supplier
UUAA	UAV USS authentication and authorization
VPLMN	Visiting PLMN
W-5GAN	Wireline 5G access network
W-AGF	Wireline access gateway function
W-CP	Wireline access control plane protocol
WLAN	Wireless LAN
XRES	Expected response

About the Editor

Anand Prasad is a Partner at Deloitte Tohmatsu Cyber (DTCY) for multinational cybersecurity services and connectivity security. With over 25 years of experience in all aspects of information and cybersecurity, Anand advocates risk based holistic security approach. He was Board of Director and chairman of Board Cybersecurity Committee at the world's first 5G wholesale network, interim Chief Information Security Officer (CISO) of the leading financial firm of Japan, founder and CEO of a startup (acquired by DTCY), a global provider of cybersecurity services and consulting with top-tier clients right across the telecommunications industry, and Senior Security Advisor of the leading telecom operator in Japan, providing advise on all aspects of information and cybersecurity. Anand was also CISO and Board Member, of the world's leading MNO with the very first cloud-native 4G and 5G network implementation. As CISO, Anand led all aspects of enterprise and mobile network security from design and deployment to operations. Anand has also held key roles in several leading multinational technology firms with roles around the globe. He was Chairman of 3GPP SA3 where, among others, he led the standardization of 5G. He is advisor to several organizations, an innovator with 50+ patents, a recognized keynote speaker and a prolific writer with 8 books and 50+ publications. He is a Fellow of IET and IETE, Editor of Cyber Security Magazine and Editor-in-Chief of the Journal of ICT Standardization published by River Publishers.

CHAPTER

1

5G Introduction

Anand R. Prasad[1] and Sivabalan Arumugam[2]

[1]Partner at Deloitte Tohmatsu Cyber (DTCY), Japan
[2]Rakuten Mobile Inc, Japan

In this chapter we present an introduction to 5G to set a background for the following chapters. We start the chapter with an explanation of factors that led to 5G standardization followed by a discussion on security architecture including the radio access network (RAN) and core network. We conclude the the chapter with a discussion on 5G security features and also cover the holistic security aspect of people, process and technology.

1.1 Factors Leading to 5G

5G has seen tremendous growth of usage around the globe in the mobile broadband arena. Although mobile broadband has seen growth, 5G was standardized to cater for a wide variety of services. The rise of digitalization, new technology availability in all spaces, the move towards cloud and increased software service based approaches were all taken into account when developing 5G.

Technology enhancements and market requirements led to 5G moving towards a cloud based distributed architecture unlike any other previous mobile communication generations. Network slicing was standardized, allowing networks to be provisioned based on service requirements. Network exposure also resulted from meeting the needs and desires of business partners that

DOI: 10.1201/9788743810902-1

have some degree of influence over the network that serves them. However, this also led to widespread usage of common web protocols and APIs. All of this naturally facilitates the development of mobile communications services, a multi-vendor ecosystem, and growing use of open source.

The 5G standard was designed to address this wide range of use cases and applications and thus the technology is highly customizable in terms of (a) the selection of radio spectrum, (b) on premise deployment and (c) IT/OT integration. Additionally, the 5G standard also defines non-public networks (NPNs) allowing various vertical services (IoT, OT, etc.). The three core focus areas of 5G are depicted in Figure 1.1:

- Enhanced mobile broadband services (eMBS), with mobile data rates much higher than 4G, as is used today often by mobile devices for video streaming, gaming, etc.
- Massive Internet of Things (mIoT) allowing a huge number of IoT devices to communicate over the mobile network where examples include light bulbs or products in containers of freight. These kinds of services could work with lower data rates but require long battery life.
- Ultra reliable low latency communication (URLLC). This is the set of IoT communications requiring higher reliability such as vehicle to vehicle communication.

Figure 1.1 5G security focus areas.

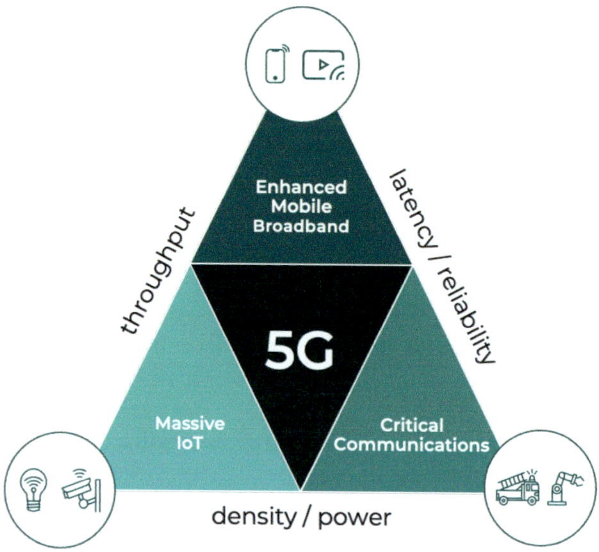

1.2 5G System Architecture

Based on the requirements in the previous section, the 5G system architecture is designed such that it not only enhances radio performance but also redefines network deployment, orchestration and service agility. 3GPP always defined modular architecture but this has been further enhanced in 5G, and we will cover the radio and core network parts in the following sections. The benefit of the architecture is that the operators can tailor deployments to specific use cases, optimize resource placement and automate lifecycle management. Additionally, 5G uses a cloud-native architecture, which naturally incorporates microservices and containerization. As a result, 5G dramatically changes mobile networks and makes DevOps feasible. As always with 3GPP, the 5G architecture also covers seamless migration from previous generations, i.e. 4G. Other benefits the 5G architecture brings are in the form of network slicing, allowing various parties to have their own share of network services, and partnership business opportunities by opening APIs to external parties.

1.2.1 NG RAN

The 5G RAN, also known as open RAN, comes with a modular base-station that can be completely distributed and is designed to benefit from the cloud. The 5G RAN, known as the next generation-RAN or NG-RAN, comprises of gNodeBs (gNBs) and can comprise of ng-eNB which is an enhanced 4G base-station. The gNB is split into a gNB-central unit (CU) and a gNB-distributed unit (DU); see Figure 1.2. The CU holds higher layer protocols while the RU has lower layer protocols and thus handles time-sensitive radio functions closer to the radio unit (RU). When it comes to deployment, one has several options such as DUs placed close to the RUs while placing the CU at a regional cloud, taking advantage of multi-access edge computing (MEC). Interfaces between all gNB functions are also already defined by 3GPP such as F1, consisting of F1-U for the user-plane and F1-C for control-plane, and an NG interface connecting to the core network. Further specification is done by various bodies such as ORAN on NG RAN and enhancement of services provisioned. O-RAN architecture and interfaces are depicted in Figure 1.2.

Figure 1.2 The 5G base-station or gNodeB (gNB).

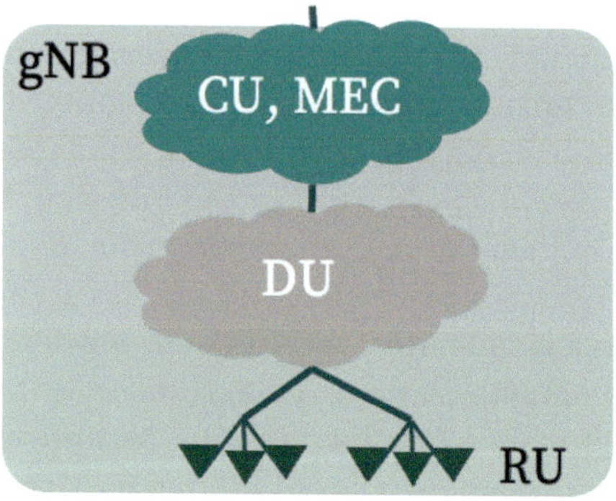

1.2.2 5G core–service based architecture

The 5G core (5GC) adopts a service based architecture (SBA) that enables the control plane nodes, see Figure 1.3, to access any services that other logical nodes provide. Given that all control plane nodes are connected by a bus-type interface, known as a service based interface (SBI), each control plane function can communicate directly with the others. This network topology contributes a lot to simplifying control plane signaling to introduce new functionalities. An SBI adopts web-based technology on HTTP/2 which enables more flexible and rapid service development. A new function, a network resource function (NRF), is used as a registry of all functions in the architecture somewhat like a domain name server (DNS).

Figure 1.3 Service based architecture.

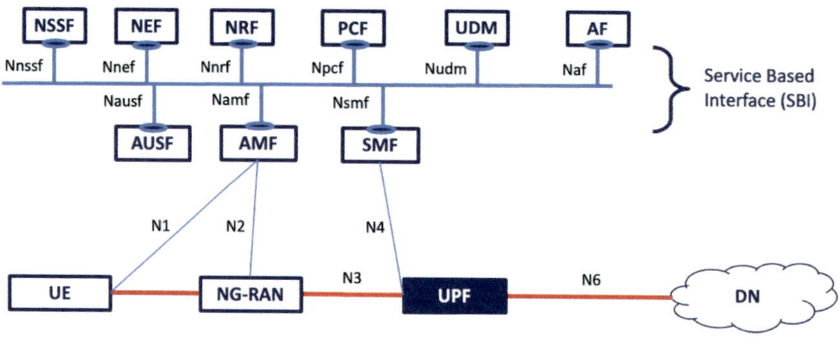

1.3 Security Features

Key security features of 5G are depicted in Figure 1.4 divided into architecture enabled, feature enhancements and novel security features; we will briefly touch on a few security features in this section.

The most important element is primary authentication, which is performed between the mobile device (user equipment or UE) and the network leading to mutual authentication. Two authentication mechanisms mandatory for 5G are the 5G authentication and key agreement (5G-AKA) that has home network control and the extensible authentication protocol-AKA' (EAP-AKA'), which enables a unified authentication framework for 3GPP and non-3GPP access technologies. Other extensible authentication protocol (EAP) based authentication methods are also optionally allowed for primary authentication in 5G. The secondary authentication is to authenticate the UE's access to the 3rd party data network (DN) via the Session Management Function (SMF), which acts as the authenticator. EAP based method is also agreed as the secondary authentication framework but the exact authentication method is left open. 5G also provides solutions for the privacy of subscriber identity by using public key cryptography. Basically, a subscriber concealed identifier (SUCI) is generated to preserve privacy from the 5G subscription permanent identifier and is called the subscription permanent identifier (SUPI).

Figure 1.4 Security features of 5G.

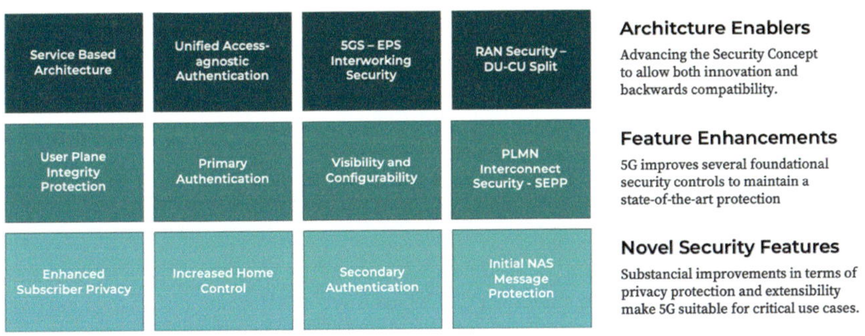

1.4 5G Security Requirements

1.4.1 Core security requirements

Typical core security requirements for a 5G network often include:

Authentication and authorization: Strong mechanisms to authenticate users, devices, and network elements, and to authorize access to network resources based on roles and permissions.
Encryption: Use of encryption to protect the confidentiality of data in transit and at rest. This includes encryption of control plane and user plane traffic.
Integrity protection: Mechanisms to ensure the integrity of data and signaling messages to prevent tampering.
Secure key management: Secure key management refers to the creation, distribution, and storage of cryptographic keys in a secure manner.
Network segmentation: Dividing the network into logical segments to limit the impact of security breaches.
Security monitoring and logging: Continuous monitoring of network activity for security threats and comprehensive logging of security events.

1.4.2 Service-based architecture (SBA)

In an SBA, security requirements often revolve around:

Secure API access: Protecting APIs used by network functions with authentication, authorization, and rate limiting.
Mutual authentication: Network functions authenticating each other to prevent unauthorized access.
Service mesh security: Using a service mesh to provide secure communication between microservices.
5G SA components and supporting infrastructure: This emphasizes a holistic approach, ensuring that security is considered throughout the entire 5G SA ecosystem.

1.4.3 Visualized deployment

Security considerations in virtualized deployment either as container network functions (CNFs) and virtual network functions (VNFs) often include:

Image scanning: Scanning container images for vulnerabilities before deployment.
Runtime security: Monitoring container activity for malicious behavior.
Isolation: Isolating containers from each other to prevent lateral movement in case of a breach.
Secure orchestration: Securing the container orchestration platform (e.g., Kubernetes).

1.4.4 Supply chain security

This is a broad area, but some common measures include:

Vendor risk management: Assessing the security practices of vendors.
Software bill of materials (SBOM): Requiring vendors to provide an SBOM to identify components and potential vulnerabilities.
Secure development practices: Ensuring that vendors follow secure development practices.
Tamper-proofing: Implementing measures to prevent tampering with hardware and software.

1.4.5 Fraud management

The mentioned GSMA recommendations (FS.21, FS.34, FS.20, FS.37, FS.19) provide detailed guidance on specific aspects of fraud prevention related to signaling, key management, and protocol security.

1.4.6 NESAS

NESAS provides a standardized framework for assessment of the security of network equipment. It involves security audits and testing to ensure that equipment meets certain security requirements defined by 3GPP.

Remember that the above mentioned requirements are still a general guidance and recommendations. Actual security requirements will vary depending on the specific deployment architecture and the operator's specific security policies.

1.5 5G Security Implications

Holistic security in 5G can only be achieved by addressing three interconnected domains: people, processes and technology. Each domain presents its own unique challenges and requirements to ensure a secure and resilient 5G ecosystem.

1.5.1 People

Effective 5G security starts with skilled and empowered personnel. Key considerations include:

- **Empowering the operations team:** Teams must be equipped to handle the shift toward cloud-native infrastructure, which demands new skills and approaches.

- **DevSecOps readiness:** A successful DevSecOps approach requires proficiency in programming and virtualization to integrate security into the development lifecycle.
- **Training for IT toolchains:** The transition from proprietary solutions to standardized IT toolchains necessitates robust training and continuous development programs.
- **Addressing the reality gap in automation:** While intelligent automation is a strategic goal, achieving it requires skilled roles such as:
  - **Cloud-native architects** to design scalable and secure deployments.
  - **Programmers** to integrate diverse systems.
  - **Data engineers and data scientists** to develop and manage automated decision-making capabilities.

1.5.2 Processes

Security must be embedded in processes that support agility, openness, and compliance:

- **Standardization and automation:** Critical for enabling new business models such as:
  - Customizable, self-service offerings (e.g., network slicing, multi-access edge computing).
  - Opening networks to third parties like private networks or over-the-top (OTT) service providers.

- **Rapid iteration and deployment:**
  - Managing dynamic workloads that scale and migrate across environments.
  - Ensuring continuous deployment of updates and security patches.

- **Policy-driven management:**
  - Leveraging declarative models where users define high-level policies.
  - Orchestration systems translate policies into specific network configurations.

- **Runtime compliance and validation:**
  - Ensuring real-time adherence to industry standards and regulatory requirements through continuous validation.

1.5.3 Technology

Technological complexity in 5G introduces new vectors for security challenges:

- **Securing the supply chain:**
  - Managing security in increasingly heterogeneous and multi-vendor deployments.

  - Dealing with both open-source components and proprietary software.

- **Asset and identity management:**
  - Tracking assets and managing identities and credentials throughout:
  - Workload instantiation.
  - Secure provisioning and storage.
  - Lifecycle cleanup and decommissioning.

- **Workload monitoring and auditing:**
  - Handling the ephemeral nature of cloud-native workloads.
  - Managing workloads that migrate within the network or are owned by external parties (e.g., MEC tenants, slice owners).

Bibliography

[1] 3GPP TS 23.501 V16.2.0, "System Architecture for the 5G System (5GS); Stage 2," Dec 2019.

[2] 3GPP TS 33.501 V16.1.0, "Security architecture and procedures for 5G System," June 2019.

[3] Prasad, A. R., Mademann, F., & Guttman, E. "Special issue on 3GPP 5G Specifications" Journal of ICT Standardization, 6(1-2), River Publishers, 2018.

[4] ETSI GS NFV-SEC 003 V1.1.1, "Security; Security and Trust Guidance," Oct 2019.

[5] GSMA, "Securing the 5G Era," Mobile Security and Fraud Forum, Feb 2020.

[6] NGMN Alliance, "5G Security Recommendations Package #2," July 2020.

[7] O-RAN Alliance, "O-RAN Architecture Description v3.0," July 2021.

[8] ONAP, "ONAP Architecture Overview," The Linux Foundation, 2022.

[9] ITU-T Y.3101, "Requirements of the IMT-2020 network," 2020.

[10] Prasad, A. R., Arumugam, S., Baskaran, S. B. M., & Zugenmaier, A. (2018). "3GPP 5G Security." *Journal of ICT Standardization*, 6(1-2), 137–158

[11] O-RAN Architecture Description 15.0- O-RAN.WG1.TS.OAD-R004-v15.00 October 2025 Technical Specification R004

CHAPTER

2

5G Authentication Mechanisms

Takahito Yoshizawa[1,*], Sheeba Backia[2], Mary Baskaran[2], and Xiaoting Huang[3]

[1]COSIC, KU Leuven, Kasteelpark Arenberg 10 Bus 2452, B-3001 Leuven, Belgium
[2]Lenovo, Motorola Mobility, Frankenweg 23, 61381 Friedrichsdorf, Hessen, Germany
[3]China Mobile Research Institute, No.32 Xuanwumen West Street, Xicheng District, Beijing, China
*Corresponding author

Email addresses: takahito.yoshizawa@esat.kuleuven.be; smary@lenovo.com; huangxiaoting@chinamobile.com

Abstract

Since the commercialization of 1G systems in the 1980s, mobile systems have evolved over the course of generations. Since the introduction of packet switched data services in 3G and the emergence of so-called "smartphones," the use of mobile phones has started to shift from voice calls to data-oriented usages. With the introduction of the 4G system, data traffic has become predominant. Fostered by the convergence with a flat IP-based data network, data-oriented mobile network usage expanded to different types of devices and usages such as UAVs, IoTs and machine-type communication (MTC) in addition to human users. Commercial deployment of 5G systems started in 2020 in various parts of the world, and is now the dominant technology for various usages,

DOI: 10.1201/9788743810902-2

industries, and applications. As the systems and services evolve, so does the security landscape of mobile systems. This situation drove the security solution of mobile systems to evolve and adapt in order to address security concerns associated with the distributed network's new services and usages. Further technological and architectural evolutions, such as the use of virtualization and service-based architecture (SBA) in 5G, also introduced new types of user authentication. User and device authentication is no longer a single procedure in the 5G system. In this chapter, we discuss various types of authentication mechanisms in the 5G system based on the Release-18 specifications.

2.1 Introduction

3GPP has specified the system and security architecture for the 5G system (5GS). The first 5G specifications were published around the end of 2017 to the beginning of 2018 as the Release 15 version. Release 15 is called 5G phase I, which specifies the initial set of features to launch the commercial 5G deployment. Under Release 16, called 5G phase II, 3GPP continues the 5G specification work by introducing enhancements and additional new features. 3GPP is currently at work on Release 18 and has launched the development of Release 19 in its specifications, which relate to 5G-Advanced. The latest versions of the 5G system architecture, procedures and security architecture are defined in TS 23.501 [1], TS 23.502 [2] and TS 33.501 [3].

In the 5G system, a number of security enhancement features are introduced to make the mobile system even more secure and robust compared to the preceding generation systems. The 5G version of the authentication mechanism is one example. Primary authentication offers two mechanisms: (1) the 5G authentication and key agreement (5G AKA), and (2) the extensible authentication protocol AKA' (EAP-AKA'). They address existing security issues in earlier generation systems. Specifically, introducing "home control", i.e., authentication to take place in the home network, and the concept of the subscription concealed identifier (SUCI) in the signaling prevents the exposure of the permanent subscriber ID to ensure user privacy: subscription permanent identifier (SUPI), which can have also the format of a traditional international mobile subscription identifier (IMSI). Secondary authentication provides a mechanism for the external network to authenticate the user with the support of a mobile operator.

In addition, the 5G system introduces several new concepts that did not exist in preceding generation systems. One such area is the concept of the network slice, which was originally proposed by the Next Generation Mobile Networks

(NGMN) Alliance as one of the key concepts in the 5G system. It allows mobile operators to deploy emerging technologies such as network virtualization based on software defined network/network function virtualization (SDN/NFV). Network slicing provides network isolation tailored to fulfil diverse service requirements for various applications of vertical industries. The introduction of the network slice concept also enables authentication at the network slice level. Finally, Release 16 introduces "authentication and key management for applications based on 3GPP credentials in 5G" (AKMA). This new bootstrapping architecture enhances the existing generic bootstrapping architecture (GBA)-based solution of the 4G and 3G systems.

While developing the above features, it was found there was no mechanism for the home network to trigger primary authentication which is required by several use cases in order to refresh K_{AUSF}. The home network uses K_{AUSF} or keys derived from K_{AUSF} to provide protection for various services, (e.g. interworking from LTE to 5G, steering of roaming/UE parameter update (SoR/UPU) and AKMA services). Therefore, 3GPP has introduced the capability of the home network triggered primary authentication since Rel-18; with that, the home network would benefit from having the ability to be able to ensure a fresh K_{AUSF} is available by triggering an authentication, in particular to prevent counter wrap in SoR/UPU or after interworking from LTE when there might be no K_{AUSF} available.

The 5G system is designed as access-agnostic, where not only the new radio air but also a non-3GPP access network, e.g., WiFi, is enabled to get access to the 5G services. Since Rel-16, 3GPP has introduced the concept of trusted and untrusted non-3GPP access and wireless wireline convergence. Security for trusted non-3GPP access to the 5G core network is achieved when the UE registers to the 5GC via the trusted non-3GPP access network (TNAN). The UE registers to 5GC and, at the same time, it authenticates with the TNAN by using the EAP-5G procedure, similar to the one used with the registration procedure for untrusted non-3GPP access.

To support wireless and wireline convergence (WWC) for the 5G system, two new network entities, 5G residential gateway (5G-RG) and fixed network RG (FN-RG), are introduced [1]. The 5G-RG acts as 5G UE and can connect to 5GC via wireline access network (W-5GAN) or via fixed wireless access (FWA). The 5G-RG also acts as an end point of N1 and provides the NAS signaling connection to the 5GC on behalf of the AUN3 (authenticable non-3GPP) devices behind the 5G-RG. The FN-RG can connect to 5GC via a wireline access network (W-5GAN).

Since Rel-16, 3GPP has started to introduce a non-public network (NPN) which is kind of private network that exclusively offers mobile network services for a specific or dedicated set of subscribers or network elements belonging to an organization (e.g., any production plant, campus) to enable private usages. NPN can be deployed either as a standalone network (SNPN) or as part of a public network, i.e., public network integrated NPN (PNI-NPN). One of the major differences of non-public networks is that authentication methods other than AKA based ones may be used in a standalone non-public network (SNPN).

To support uncrewed aerial systems (UAS) connectivity, 3GPP has specified the UAV USS authentication and authorization (UUAA) procedures to ensure that the UAV can be authenticated and authorized by a USS before the connectivity for UAS services is enabled and a UAV is allowed to perform UUAA with the USS/UTM only after the UAV (UE) has completed successfully primary authentication.

Due to the introduction of these additional features, the 5G system has various authentication mechanisms – all of them are intended for different purposes. The goal of this paper is to clarify the differences and usages of these authentication mechanisms in the 5G system.

This paper is organized as follows. We first present the overall 5G security architecture in Section 2.1.1. In Sections 2.2 and 2.3, we describe the details of primary and secondary authentication, respectively. Then, we further discuss slice authentication and AKMA in Sections 2.4 and 2.5, followed by the introduction of home network triggered authentication in Section 2.6. We further give details of support of non-3GPP access and wireline access in Sections 2.7 and 2.8, with the introduction of non-public network (NPN) authentication and UUAA given in Sections 2.9 and 2.10. Finally, we conclude our discussion in Section 2.11.

2.1.1 Overview of authentication and security of the 5G system

The 5G system architecture [1] was specified to support the new principle of the service based architecture (SBA). It allows scalable deployment of virtualized network functions (NF) that enables operators to deploy it in data centers. The idea is to shift from the hardware "box" paradigm where individual functional entities are separately defined to a model, in which NFs follow the web-based approach using RESTful client server communication. An NF as a service consumer can subscribe to certain events from another NF that acts as a service producer of the requested service information. On the other hand, the 5G system architecture can also be depicted in a traditional reference model and not only in the form of the new SBA reference model. For the security architecture, this type of visualization is chosen in order to show the co-location of functions with others more easily.

Figure 2.1 5G security architecture.

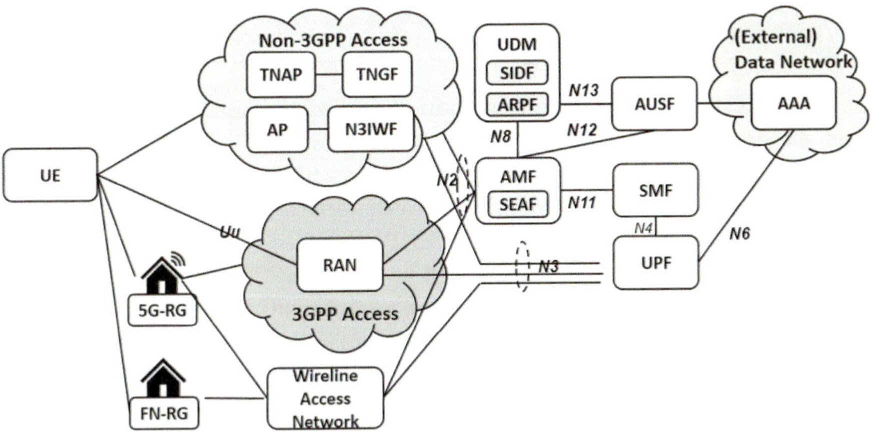

Figure 2.1 shows the overall 5G system architecture and its NFs involved in the different security procedures, especially for authentication. The user equipment (UE) is the mobile device comprising the mobile equipment (ME) and the universal subscriber identity module (USIM), which securely stores the shared credentials with the mobile operator. The UE can connect to the mobile core network via non-3GPP access, e.g., WLAN, or via 3GPP access. Untrusted non-3GPP access is shown here as an example.

The untrusted non-3GPP access consists of the access point (AP) and the Non-3GPP Interworking Function (N3IWF). The 3GPP Access consists of the Radio Access Network (RAN), gNB, i.e. the 5G base station. The UE connects to the Access and Mobility Management Function (AMF) over the Non-Access Stratum (NAS) protocol either via non-3GPP Access or 3GPP Access. The AMF hosts the SEcurity Anchor Function (SEAF). The Authentication Server Function (AUSF) is the peer point of the UE for the primary authentication. The Unified Data Management (UDM) hosts the subscription data in the Unified Data Repository (UDR) and also has additional functionality such as the Subscription Identifier De-concealing Function (SIDF) and Authentication credential Repository and Processing Function (ARPF). For user plane management, the Session Management Function (SMF) is used to interact with the User Plane Function (UPF) to connect to a Data Network (DN), e.g., the Internet. The Authentication, Authorization, and Accounting (AAA) Server can be located within the mobile operator domain or can belong to an external third-party service provider.

2.2 Primary Authentication

2.2.1 Purpose of primary authentication

Primary authentication achieves mutual authentication between the UE and the operator's network. In the AKA procedure between the UE and the network, the serving network authenticates the SUPI of the UE and the UE in turn authenticates the serving network identifier through implicit key authentication. For 3GPP access, the SUPI is based on IMSI; for non-3GPP access, the SUPI is based on a network access identifier (NAI). The "implicit key authentication" is a process where the authentication is confirmed through the successful use of keys generated from the successful AKA in subsequent procedures (e.g., protecting control plane and user plane messages). To perform primary authentication for the 3GPP access, the 5G system defines EAP-AKA' and 5G AKA as mandatory methods and uses either one based on the subscription information. For non-3GPP access, a vendor specific method, EAP-5G is used between the UE and the N3IWF to encapsulate NAS messages. Between the UE and the AUSF, any of the authentication methods similar to 3GPP access can be used for authentication by the home network. The primary authentication can be initiated by the SEAF in serving network whenever a UE sends a registration request to the SEAF or during any procedure establishing a signaling connection with the UE, according to the SEAF's policy.

2.2.2 Primary authentication procedure

The primary authentication procedure based on EAP-AKA' involving AUSF and UDM service operations for 3GPP access is shown in Figure 2.2. Due to the limited space, we focus on the EAP-AKA'-based primary authentication method in this section.

A detailed discussion of the service operations is omitted to keep the procedure concise.

(1) The UE sends the registration request NAS message to the SEAF (which has the co-located AMF) in the serving network, containing either a SUCI or a 5G globally unique temporary UE identifier (5G-GUTI).

Figure 2.2 EAP-AKA' based primary authentication procedure.

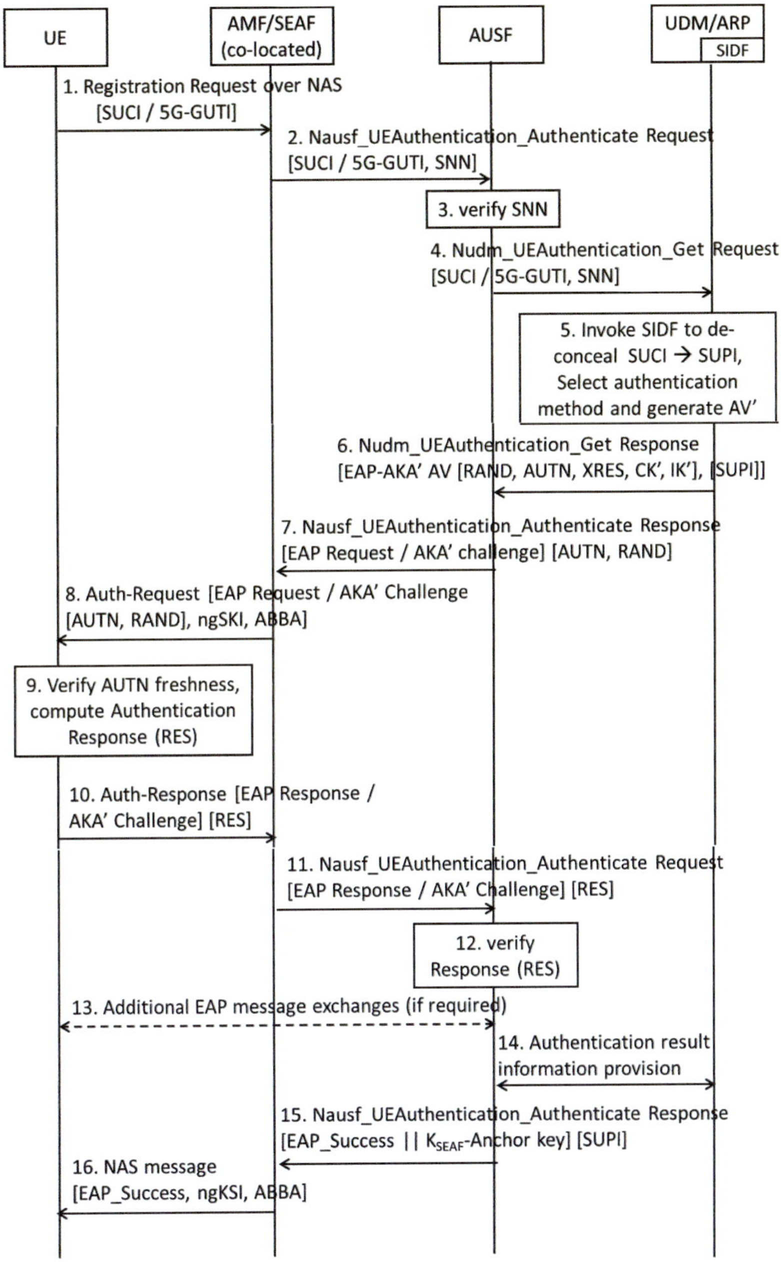

(2) On receiving the registration request, the SEAF invokes primary authentication by sending an authentication request to the AUSF in the home network (in a non-roaming scenario both serving and home network are the same) containing the received SUCI and its serving network name (SNN). Otherwise, the SEAF sends the SUPI if the SEAF has a valid 5G-GUTI and re-authenticates the UE.

(3&4) The AUSF verifies the SNN in the authentication request to check if it is the same as the expected serving name. If the AUSF verifies the serving network as authorized, it sends an authentication data request to the UDM, including the received SUCI and SNN values.

(5&6) On receiving the SUCI, the UDM invokes the SIDF to de-conceal the SUCI into the SUPI. Based on the SUPI, the UDM/ ARPF chooses the authentication method (i.e., 5G-AKA or EAP-AKA') and generates a method-specific authentication vector (AV). For EAP-AKA', a transformed AV called AV' (RAND, AUTN, XRES, CK', IK') is generated. In this discussion, we assume EAP-AKA' is selected and a transformed AV called AV' is sent in an authentication data response to the AUSF including an indication of the selected authentication method (i.e. an indicator for EAP-AKA'). If a subscriber has an AKMA subscription, the UDM includes the AKMA indication and routing indicator in the Nudm_UEAuthentication_Get Response, further details of the AKMA are introduced in Section 2.5.

(7) The AUSF sends the authentication challenge (RAND and AUTN) from the AV' to the SEAF in the *authentication response* message.

(8) The SEAF forwards the received authentication challenge to the UE including the key set identifier (ngKSI) and the anti-bidding down between architectures (ABBA) parameters in the *authentication request* NAS message. The ngKSI and the ABBA are to identify the key set and for anti-bidding down protection in the 5GS [3], respectively.

(9&10) At the UE, on receiving the RAND and AUTN, the USIM verifies the freshness of the AV by checking whether the AUTN is valid (TS 33.102) [4]. If the AUTN is valid, the UE computes a response (RES) and the keys (CK/IK) or (CK'/IK' for EAP-AKA'). The UE then responds to the SEAF by sending an *Authentication response* NAS message including the RES.

(11) The SEAF forwards challenge response transparently to the AUSF in an *authentication request* message.

(12–14) The AUSF verifies the challenge response and if the verification is successful, the AUSF informs the UDM about the authentication result as "success." For the EAP-AKA' method, there may be additional EAP message exchanges between the UE and the AUSF.

(15) The AUSF derives an extended master session key (EMSK) from CK' and IK' (RFC 5448 [5]) and uses the most significant 256 bits of EMSK as the key KAUSF, then calculates the key K_{SEAF} (the anchor key from which the AS and NAS protection keys are derived) from K_{AUSF}.

(16) The SEAF further generates K_{AMF} from K_{SEAF} received from the AUSF. If the AUSF and the SEAF confirms a successful authentication, then the SEAF provides the ngKSI and the K_{AMF} to the AMF. The SEAF further sends the *EAP success* message to the UE along with the ngKSI and the ABBA parameters.

On receiving the *authentication success* message, the UE generates the security context equivalent to the one in the network. The use of security context for the control plane and the user plane message protection between the UE and the network ensures successful completion of the authentication between the network and the UE.

2.3 Secondary Authentication

2.3.1 Principle of secondary authentication

Secondary authentication occurs between the UE and the DN outside the mobile operator domain. In the older generation systems, DNs conduct the access control by themselves without the support of a mobile operator after the user plane tunnel has been established between the UE and the DN. This may allow malicious UEs to invoke an authentication service provided by the DN resulting in a denial of service (DoS) attack. The 5G system allows mobile operators to delegate the authentication and authorization to a third party hosting the DN. This is achieved by introducing the concept of secondary authentication. This procedure is executed during the establishment of user plane connection after the successful primary authentication.

One of the delicate designs of secondary authentication is the use of an EAP framework, which is widely used and enables various credential types and authentication methods used by different application service providers.

2.3.2 Secondary authentication procedure

The prerequisite of secondary authentication is the successful completion of primary authentication based on the UE's 5G network access credentials, and the establishment of an NAS security context between the UE and the AMF. Additionally, the UE needs to be provisioned with the external credentials used for the authentication between itself and corresponding data network authentication authorization and accounting (DN-AAA) server.

Figure 2.3 illustrates the secondary authentication procedure. During this procedure, the SMF in the 5G core network performs the role of EAP authenticator while the DN-AAA server performs the role of EAP authentication server.

Figure 2.3 Secondary authentication procedure.

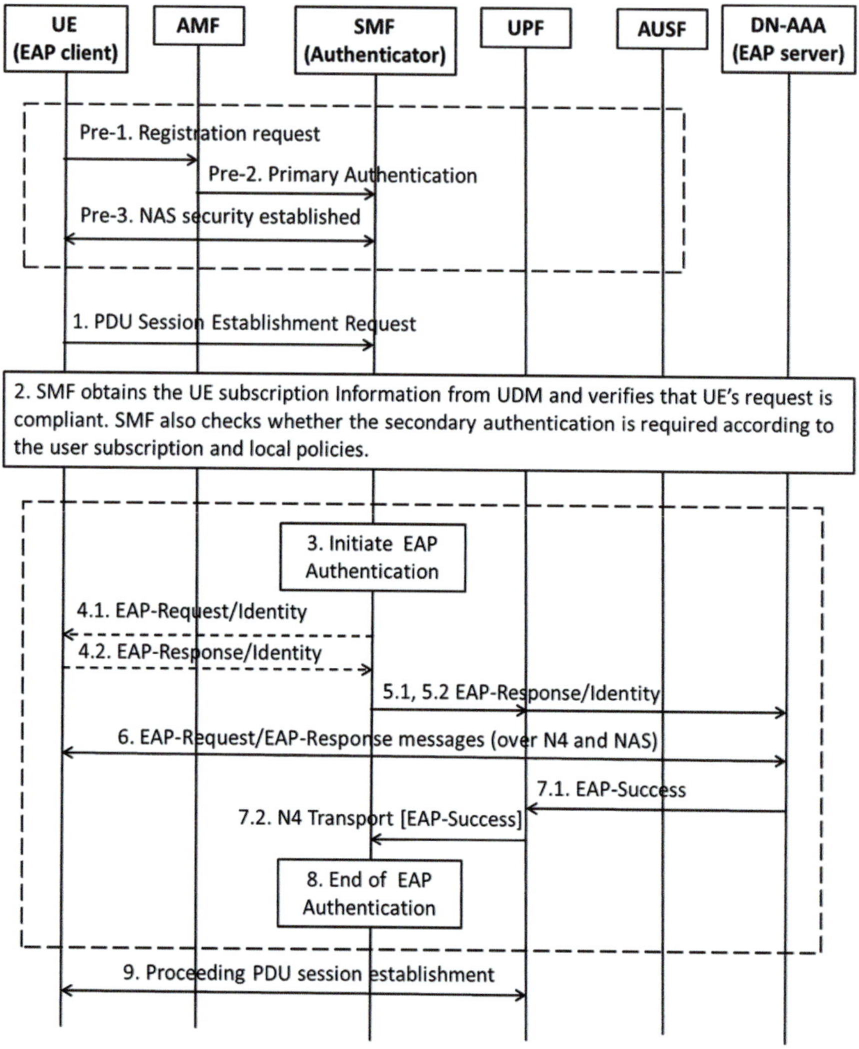

(1) The UE requests to establish a new PDU session; *PDU session establishment request* message is contained in an NAS message to the SMF.

(2) If the DN-specific identity, data network name (DNN), is provided by the UE in step 1, and the SMF determines that authentication/authorization of the PDU session establishment is required based on the SMF policy associated with the DN, the SMF triggers the secondary authentication procedure with the UE.

(3) The SMF triggers EAP authentication to obtain authorization from an external DN-AAA server.
(4) The SMF requests the EAP identity from the UE, and the UE responds with EAP identity (note: the EAP identity may also be sent in step 1 to avoid interactions in this step).
(5) The SMF forwards the *EAP response/identity* message to the DN-AAA server via the UPF.
(6) The DN-AAA server and the UE exchange EAP messages, as required by the EAP method.
(7) Once the authentication is successfully completed, the DN -AAA server sends an *EAP success* message to the SMF via the UPF.
(8) The SMF completes the EAP authentication procedure and saves the authentication results of the specific UE and the DN.
(9) If the EAP authentication procedure is successful, PDU session establishment proceeds further.

The DN-AAA server or the SMF may initiate the re-authentication with the UE. If the re-authentication is initiated by the DN-AAA server, the UE is addressed by the generic public subscription identifier (GPSI), which is notified to the DN-AAA server at step 3.

2.3.3 Comparison of primary and secondary authentication

The purposes and characteristics of the primary and secondary authentication can be summarized as follows:

(1) Primary authentication provides the access control to the operator network, while secondary authentication enables the UE to access the user plane data towards the DN.
(2) Primary authentication is a mandatory procedure when the UE connects to the operator network. On the other hand, secondary authentication is optional to use depending on the agreement between operators and the external service provider.
(3) Primary authentication uses 3GPP credentials while secondary uses external credentials provided by third-party application service providers.

2.4 Slice Authentication

2.4.1 Purpose of slice authentication

3GPP defines a network slice as a logical network resource that provides specific network capabilities and characteristics. The slice authentication, as specified in TS 33.501 clause 6.2 [3], enables the third-party application service provider hosted with an AAA server to authenticate the end-user to access its slice. A third-party application service provider offers a communication service to end-users using the service built with a network slice over the operator network, e.g. a taxi company using a network slice optimized to dispatch and manage their vehicles by tracing vehicle location and providing an uncongested route for the vehicle. Slice authentication enables customized authentication of UEs for slice selection and access to it. In addition to primary authentication, slice authentication is executed at the time of the registration when access control to network slices requires slice specific additional authorization and authentication. In this case, a slice specific user identity (e.g., NAI of the form user@domain) and a third-party provisioned credential different from the 3GPP credential in primary authentication is used. Without slice authentication, unauthorized UEs may consume resources of the network slice leading to DoS for other legitimate UEs.

2.4.2 Slice authentication procedure

The UE needs to be provisioned with a slice-specific credential necessary to authenticate itself with the AAA server. The assumption is that the slice authentication is performed for specific single network slice selection assistance information (S-NSSAI) only after the primary authentication is successfully completed. This procedure is shown in Figure 2.4. If multiple slices are requested by the UE and some of them require slice-specific authentication, then the authentication procedure is performed for each of those S-NSSAIs.

During the primary authentication, as discussed in Section 2.3.2, the UE sends a registration request message to the network containing a list of NSSAIs corresponding to the network slices and indicates if it has the capability to support slice authentication. The UE and the network complete the primary authentication either using the 5G AKA or EAP-AKA' procedure, as discussed in Section 2.3. After the successful primary authentication, the AMF sends a registration accept message to the UE with a list of all allowed NSSAIs (except the S-NSSAIs, which require slice-specific authentication) based on the subscription information available in the UDM. The UE sends a *registration complete* message, indicating the successful completion of the primary authentication.

Then the AMF initiates slice specific authentication using the procedure shown in Figure 2.4 for all slices that require slice-specific authentication.

(1&2) If the AMF determines that slice-specific authentication is required, the AMF acts as the EAP authenticator and sends an *EAP identity request* message to the UE to initiate slice specific authentication corresponding to the NSSAI of the network slice. The UE provides the EAP ID by sending the *EAP identity response* message to the AMF for the requested S-NSSAI. The AMF forwards the EAP ID along with the S-NSSAI to the NSSAAF.

Figure 2.4 Slice authentication procedure.

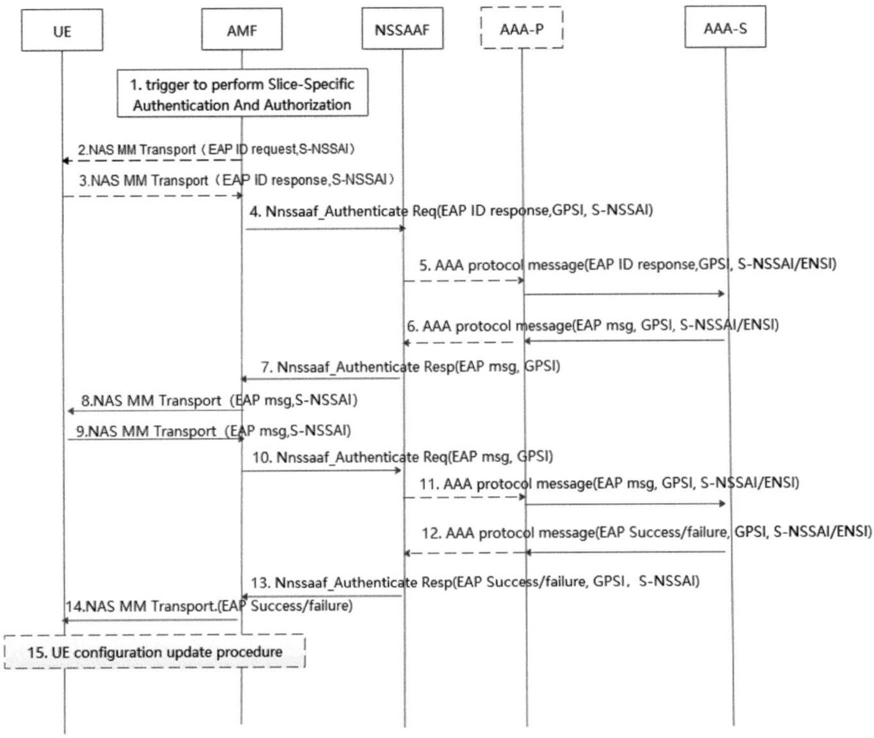

The NSSAAF forwards the *EAP identity response* message in the AAA protocol message to the serving AAA (AAA-S) server. A proxy-AAA server (AAA-P) may be present if the AAA-S is located in the third party.

(6–11) The AAA-S and the UE exchanges EAP messages.

(12) EAP authentication completes. The AAA-S delivers either an *EAP success* or failure message to the NSSAAF.

(13&14) The NSSAAF forwards the *EAP success/failure* message to the AMF in the Nssaaf NSSAA *authenticate response* message, and the AMF forwards it over MM as the NAS *transport* message.

(15) Once the slice-specific authentication is successfully completed for all S-NSSAIs, the AMF triggers a UE configuration update procedure to deliver a new list of allowed NSSAIs to the UE reflecting the slice authentication result.

In addition, the slice authentication supports two additional procedures: (1) slice-specific re-authentication and re-authorization and (2) slice-specific authorization revocation. These procedures allow the AAA-S to either re-authenticate/re-authorize or revoke the access to a specific slice that has already been granted to the UE as needed. If AAA-P is present, then it routes the message to the serving AMF based on the binding between the User ID and the GPSI of the UE established when the UE was authorized for the slice.

If the AAA-S belongs to the third party, the NSSAAF optionally maps the S-NSSAI to external network slice information (ENSI) and forwards the *EAP identity* message to the AAA-S together with ENSI and GPSI. In this case, the AAA-S uses the EAP-ID and ENSI to identify the UE for which slice authorization is requested.

2.5 AKMA

2.5.1 Purpose of AKMA

5G promises to enable massive Internet of Things (mIoT) applications. When getting access to the IoT application server, IoT devices may have limitations in applying username/password authentication mechanism to their application servers, which is common in Internet style authentication mechanisms. An IoT device may lack resources in storing and/or processing certificates or may not have provisioning interfaces such as a graphical user interface (GUI). Therefore, a new authentication scheme applied to IoT devices and their application servers is needed to allow only the authentic devices to access application servers. Besides, the question of how to enable secure communication between UE and application providers draws attention.

In preceding generation systems, 3GPP has developed well-designed mechanisms, e.g., GBA [7] and battery efficient security for very low throughput machine type communication (MTC) devices (BEST) [8] that leverage operators' assets, namely using 3GPP credentials, to achieve authentication and key distribution for application providers. However, two gaps need to be filled to fit these mechanisms into the 5G system. First, the 5G architecture introduces the concept of SBA and new network functions different from the preceding generation systems. Second, the variation of IoT application layer protocols, which are not supported by both GBA and BEST, needs to be supported.

To solve the above-mentioned problems, 3GPP has studied [9] and specified [10] a new feature in Release 16, which is called AKMA (authentication and key management for applications). The goal of this feature is to enable the authentication and generation of application keys based on 3GPP credential for all UE types in the 5G system, especially IoT devices, ensuring bootstrapping the security between the UE and the applications in the 5G system.

2.5.2 AKMA procedure

Since AKMA is intended to leverage the 5G authentication mechanism to provide applications with session keys that are afterwards used to protect the communication between the UE and the application servers, an anchor function is needed to act as the role of generating and managing such session keys. In 3GPP TS 33.535 [10], the AKMA anchor function (called AAnF) has been defined in the 5G system to obtain the intermediate AKMA key (K_{AKMA}) from the AUSF, as well as to provide a session key (K_{AF}) to application providers. Considering the 5G security design of enabling KAUSF storage in the home network and the UE for further services, AKMA is designed to leverage primary authentication and namely K_{AUSF} to simplify the procedures. Specifically, if the UE is subscribed with AKMA service capability, there will be a K_{AUSF} stored and a further K_{AKMA} derived from the K_{AUSF} after the successful primary authentication (Figure 2.5).

(1&2) During the primary authentication, the AUSF interacts with the UDM to fetch authentication information. If the UE is capable and allowed to use the AKMA service, the UDM indicates to the AUSF to generate AKMA related keys with the parameter of AKMA indication (AKMA Ind) which is stored in the UDM as subscription data. If the AKMA indication is included, the UDM also includes the routing indicator (RID) of the UE for further use of AAnF selection.

(3) While receiving the AKMA indication from the UDM, the AUSF and UE both generates K_{AKMA} from K_{AUSF} and the related key identifier A-KID (AKMA key identifier).

(4&5) The AUSF sends the AKMA key material (A-KID and K_{AKMA}) together with SUPI to the AAnF for further key derivation.

(6) When the UE is about to communicate with the application server (AKMA AF), it sends an *application session establishment request* message carrying A-KID.

(7) If the AF does not have the key material associated with the A-KID, then the AF requests the K_{AF} (derived from K_{AKMA}) from the AAnF by carrying the A-KID and the AF identifier (AF Id).

(8&9) Once the request has been received from the AF, if the AAnF determines this specific AF needs GPSI, according to its local policy, the AAnF sends a Nudm_SDM_Get Request to the UDM to fetch the GPSI of the UE and UDM responds with the GPSI of the UE afterwards. If the specific AF does not need GPSI, the AAnF continues with step 10.

(10&11) The AAnF derives K_{AF} from KAKMA and sends it to AF with its lifetime, together with the UE's SUPI or GPSI based on AAnF's local policy. In some scenarios, anonymous user access to the AF is desirable (e.g., UE identification is not required at the AF), in this case, SUPI/GPSI is not returned by the AAnF.

(12) After obtaining the K_{AF}, the AKMA AF responds to the UE's application session establishment request. Afterwards, the UE derives the K_{AF} accordingly and both the UE and AF are in possession of the K_{AF} for further use.

Figure 2.5. AKMA procedure.

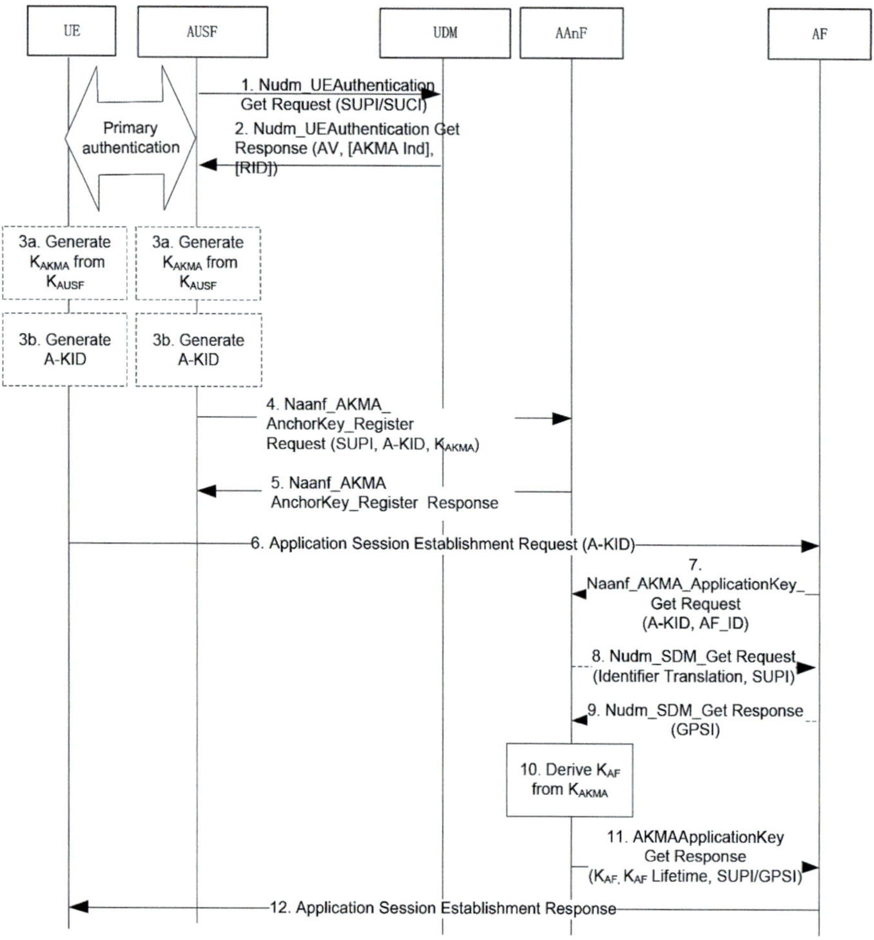

The above procedures are applicable where the AF is located inside the operator's domain. When the AF is located outside the operator's network, the AF requests K_{AF} from the AAnF via NEF (Figure 2.6). Detailed procedures are specified in 3GPP TS 33.535 [10].

Figure 2.6 AKMA key hierarchy.

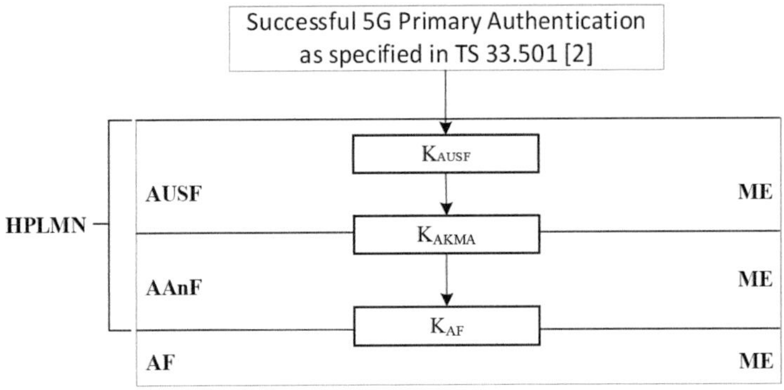

2.6 Home-network-triggered Authentication

The notable enhancement in the primary authentication mechanism in the 5G system over the previous generation systems is the additional controls by the home network. This includes the termination of the authentication procedure in the home network and the use of SUPI to encrypt the UE's permanent ID (i.e., IMSI or NAI). On the other hand, the triggering of the primary authentication procedure is still left at the visiting network in the case of a roaming scenario. In this sense, further enhancements made it possible to allow the home network to trigger re-authentication of a UE when it is deemed necessary based on an operator's authentication policy.

During the 5G primary authentication, as discussed in Section 2.0, a key hierarchy is derived for a UE as shown in Figure 2.7. As shown in this figure, K_{AUSF} is the intermediate key from which the remainder of the keys in this hierarchy is derived. There are several 5G security features that rely on K_{AUSF} such as steering of roaming (SoR) security, UE parameter update (UPU) security. Furthermore, K_{AUSF} is also used to derive keys (K_{AKMA} and K_{AF}) for other services such as AKMA discussed in Section 2.0.

The presence of this 5G key hierarchy is not always guaranteed; in the interworking scenario where a handover from LTE to 5G occurs, and this is the first handover to the 5G system, the key hierarchy is not established in the 5G CN and the UE given that the 5G primary authentication has not been executed with this UE. In other words, KAUSF is not established for this UE. Under this circumstance, any procedure that requires a valid KAUSF will fail. Another circumstance where primary (re-)authentication needs to be triggered for a UE is when the counter wraparound for the SoR/UPU is about to occur or the AKMA key lifetime is about to expire. Because the 5G primary authentication procedure is the only one that generates or refreshes the KAUSF key, it follows that the home network needs to be able to trigger re-authentication of the UE in order to generate or refresh the KAUSF key.

3GPP conducted a study of this issue and the result is summarized in TR 33.741 [13]. The subsequent normative specification was added to TS 33.501 clause 6.1.5 [3] as a part of the Rel-18 feature.

Figure 2.7 Key hierarchy as the result of 5G primary authentication.

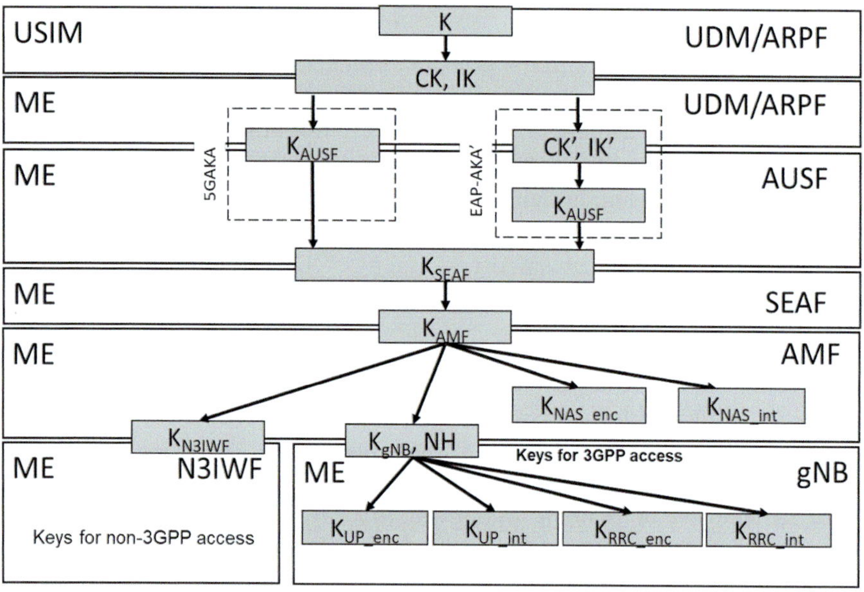

The UDM in the home network is responsible for determining and triggering the UE (re-)authentication procedure to the UE. It is based on a pre-configured operator policy in the home network. Triggering events include, e.g., UE-initiated events (UE registration to 5GC), UE-targeted events (SoR, UPU), or other NE-triggered events.

If the UDM determines triggering (re-)authentication of the UE, it sends a request to the AMF. The AMF then initiates the primary authentication procedure with this UE. As a result, the key hierarchy from KAUSF key downward is re-established.

The messages in the procedure for the home network to trigger primary authentication are confidentiality-protected, integrity-protected and anti-replay-protected.

2.7 Support of Non-3GPP Access

2.7.1 Introduction

Non-3GPP access to the mobile network is classified into two types: trusted and untrusted. Whether a non-3GPP access is trusted or not is defined by the individual mobile network operator's policy; the characteristics of the access network itself is not associated with this classification. One possible way to define the trusted non-3GPP access network is to pre-configure it in the USIM. This way, the UE can prioritize the trusted non-3GPP access network over untrusted ones. The layer-2 connection between the UE and the non-3GPP access may include technologies such as 802.11 WLAN, PPP, and Ethernet.

When a UE accesses the mobile network using a non-3GPP access network, then all of its connections to DNs are categorized as either one of the two trust relationships (i.e., trusted or untrusted). In other words, it is not possible to mix this trust relationship when the UE connects to multiple DNs over the same non-3GPP access network.

2.7.2 Untrusted non-3GPP access

In the untrusted non-3GPP access, IPsec is established between the UE and the N3IWF using the IKEv2 procedure as defined in RFC 7296. In this case, the UE plays the role of the client and the N3IWF takes the server role, respectively, to establish IPsec ESP security associations between them. EAP-5G is used as the authentication protocol between the UE and the N3IWF. N2 NAS signaling messages are encapsulated within the IKE messages over this EAP-5G signaling. The overall procedure is illustrated in Figures 2.8 and 2.9.

At this time, the AMF derives K_{N3IWF} from K_{AMF} and provides it to the N3IWF.

Figure 2.8 Untrusted non-3GPP access architecture.

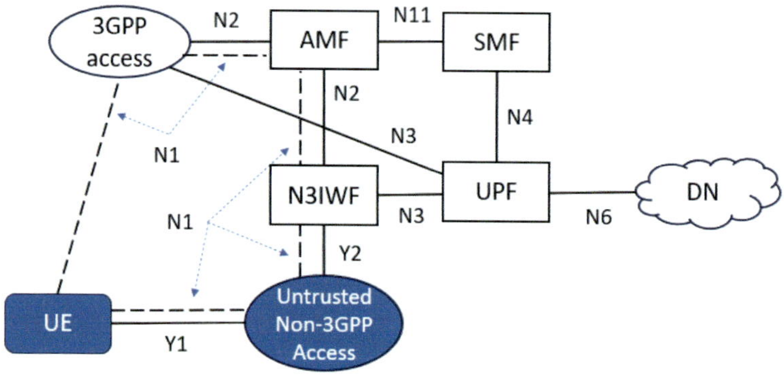

Figure 2.9 Untrusted non-3GPP access authentication procedure.

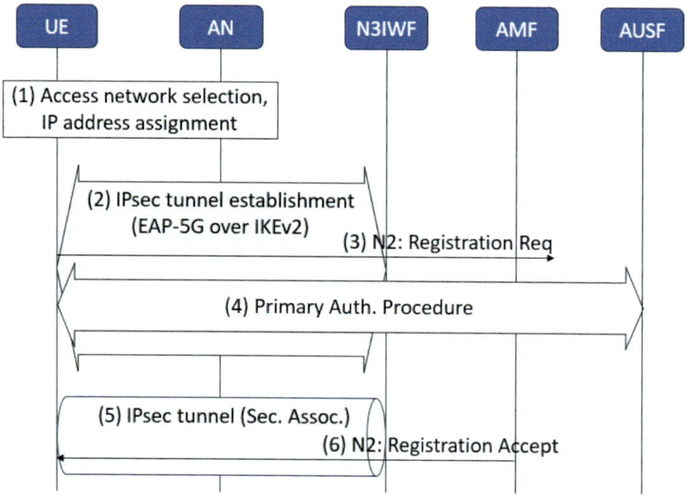

(1) First, the UE selects an access network to connect to the mobile network.

(2) Next, the UE selects an appropriate N3IWF and initiates the IPsec tunnel establishment procedure using the IKEv2 protocol. At the IKE message level, the UE uses a random number rather than its real identity such as SUCI, SUPI, or GUTI. Further, if the UE is provisioned with the root certificate of N3IWF, it requests N3IWF's certificate. In this case, the N3IWF provides its own certificate to the UE along with its identity, and the UE uses this information to verify the authenticity of the N3IWF.

(3) Upon successful verification of N3IWF, the UE initiates the EAP-5G procedure over the ongoing IKE protocol with the N3IWF. In the uplink EAP-5G message, the UE encapsulates the NAS *registration request* message to trigger the registration with the AMF. At this time, if the UE has previously established the NAS security context through the 3GPP access, thus the existing NAS security context is available, then the UE integrity-protects this *registration request* message and includes the existing 5G-GUTI instead of SUPI in this message.

(4) Upon receiving this *registration request* message, the AMF initiates the primary authentication procedure with this UE as described in Section 2.2. If there is an existing NAS security context, messages in this primary authentication procedure are confidentiality-protected and integrity-protected by using the existing NAS keys. Between the UE and the N3IWF, the primary authentication messages are encapsulated within the IKE messages.

(5) After the successful completion of the primary authentication procedure between the UE and the AMF, the IKEv2 procedure completes with successful establishment of security associations between the UE and the N3IWF.

(6) The AMF sends the NAS *registration accept* message to the UE over the IPsec tunnel, and the UE initiates a PDU session establishment.

2.7.3 Trusted non-3GPP access

In principle, the authentication procedure for trusted non-3GPP access is similar to the untrusted non-3GPP access. The key difference between them is the presence of a trusted non-3GPP access network (TNAN) in the former; it is an entity that the associated MNO trusts. TNAN consists of two entities: a trusted non-3GPP access point (TNAP) and a trusted non-3GPP gateway function (TNGF). The former serves to terminate the layer-2 connection with the UE, and the latter serves as the interworking with the 5G CN entities. The interface between TNAP and TNGF is AAA. Figure 2.10 shows the architecture of trusted non-3GPP Access and Figure 2.11 illustrates its overall procedure.

Figure 2.10 Trusted non-3GPP access architecture.

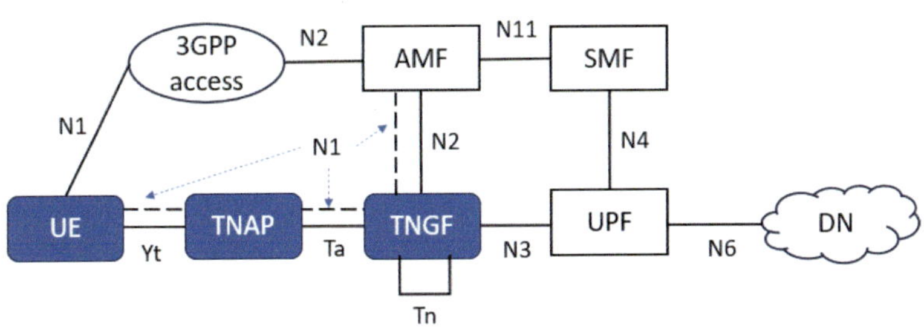

Figure 2.11 Trusted non-3GPP access authentication procedure.

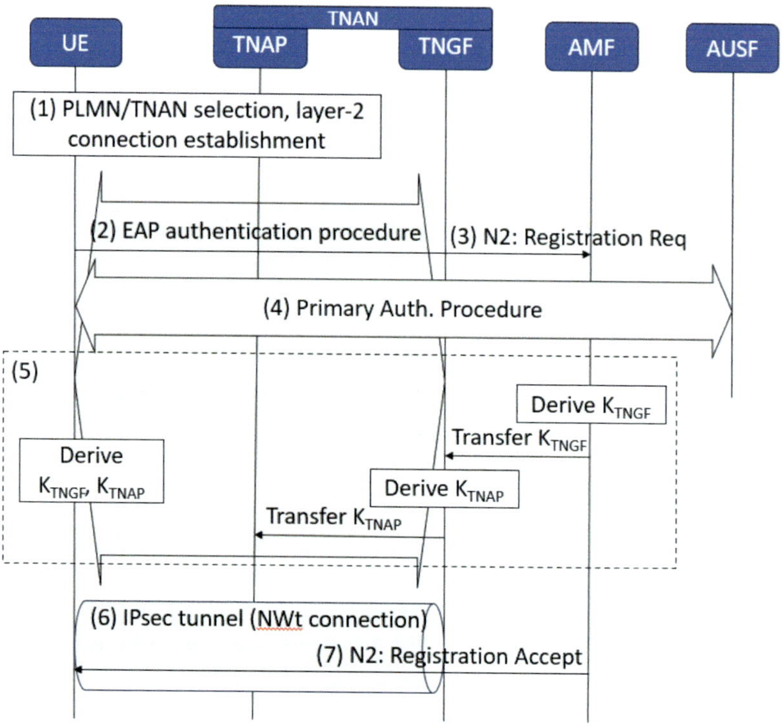

(1) The UE selects a PLMN and a TNAN for trusted non-3GPP access and establishes layer-2 connection with the TNAP. Depending on the layer-2 technology used (e.g. 802.11 WLAN), the appropriate association procedure takes place.

(2) TNAN triggers an EAP authentication procedure with the UE. EAP messages are encapsulated into the layer-2 frame being used between the UE and the TNAP. Between the TNAP and the TNGF, EAP messages are encapsulated into AAA messages.

(3) The UE initiates an EAP-5G procedure with the 5G CN via the TNAN. In the uplink EAP-5G message, the UE encapsulates an NAS *registration request* message to trigger the registration with the AMF. Upon receiving this message, the TNGF selects an appropriate AMF and forwards this NAS message.

(4) An EAP-5G procedure takes place between the UE, NTAN, and the 5G CN. Similar to step 2 above, EAP-5G messages are encapsulated in the layer-2 frame (between UE and the NTAP) and the AAA messages (between NTAP and NTGF). If the UE has previously registered to the 5GC over 3GPP access, then the NAI contains the 5G-GUTI assigned to it at that time.

(5) After successful authentication, the AMF generates K_{TNGF} and transfers it to the TNGF. This key is equivalent to K_{N3IWF} in the untrusted non-3GPP access we discussed in Section 2.0. Then the TNGF derives K_{TNAP} from K_{TNGF} and provide it to the TNAP. At the same time, the UE derives K_{TNGF} and K_{TNAP}. The TNGF provides its IP address to the UE. At this point, the EAP procedure between the UE and the TNGF completes.

This K_{TNAP} key is used to protect the traffic between the UE and the TNAP. In the case of 802.11 WLAN, this key is the pairwise master key (PMK) to execute a four-way handshake to establish the security context between the UE and the TNAP (WLAN AP) to protect traffic over the air.

(6) The UE and the TNGF establish IPsec SA. The common key (K_{TIPSec}) is used for mutual authentication. All subsequent NAS messages are transferred over this IPsec tunnel (NWt connection) between the UE and the TNGF. This tunnel uses null encryption and only integrity protection is applied. The TNGF responds to AMF with an N2 *initial context setup response* message.

(7) The AMF sends the NAS *registration accept* message to the UE over the NWt tunnel, and the UE initiates a PDU session establishment. The TNGF may establish multiple IPsec child SAs per PDU session.

Figure 2.12 shows the resulting key hierarchy for trusted non-3GPP access. As described in the procedure description, K_{TNGF} generated by the AMF is used for two purposes: (1) to IPsec tunnel between the UE and the TNGF over NWt (K_{TIPsec}) and (2) to establish session keys (e.g., WLAN keys) between the UE and the TNAP (K_{TNAP}).

Figure 2.12 Trusted non-3GPP access key hierarchy.

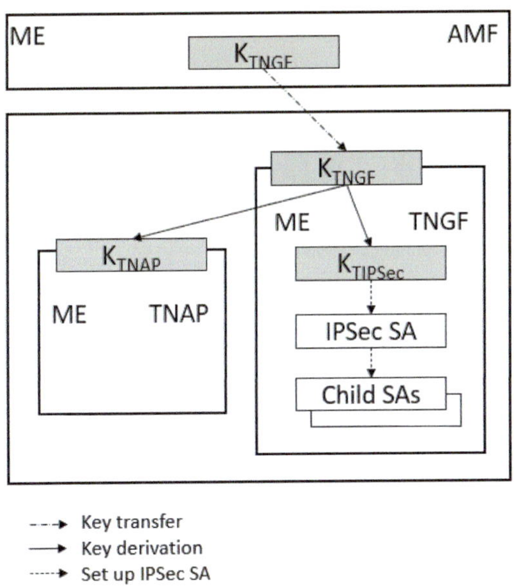

2.7.4 Non-NAS-capable device over WLAN access

The last type of non-3GPP access authentication is trusted WLAN access which is a variation of the trusted non-3GPP access discussed in Section 2.0. It is for devices that have no NAS signaling capability – non-5G-capable over WLAN (N5CW) (Figure 2.13). For to this reason, NAS messages with the 5GC needs to be terminated and processed at a "proxy" on behalf of the N5CW device itself.

Trusted WLAN access network (TWAN) provides this interworking function. It consists of two elements: (1) a trusted WLAN access point (TWAP) and (2) a trusted WLAN interworking function (TWIF). The former terminates the WLAN connection with the N5CW device and the latter terminates the NAS signaling with the 5GC and interwork with the N5CW device. Similar to the trusted non-3GPP access scenario in Section 2.0, the interface between the N5CW and the TWAP is EAP, and between the TWAP and the TWIF is AAA. Because the N5CW device does not support NAS signaling, the EAP-AKA' authentication procedure is encapsulated in the NAS authentication messages which terminate at the TWIF on behalf of the N5CW device and passed onto the N5CW over an encapsulating AAA message to the TWAP and the native EAP messages to the N5CW.

Figure 2.13 Trusted non-3GPP access architecture.

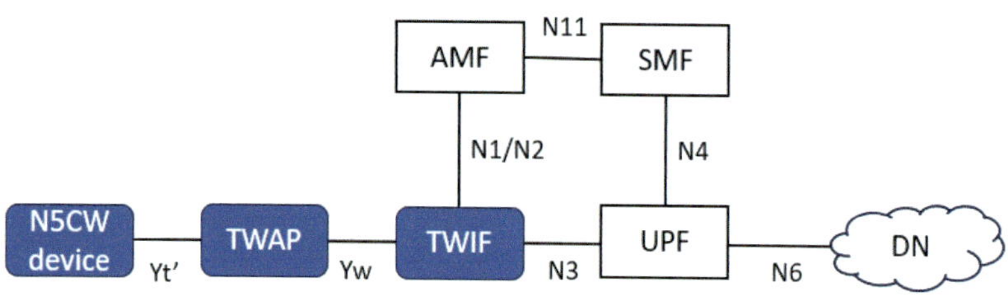

Figure 2.14 illustrates this procedure.

(1) The N5CW device selects a PLMN and a TWAN for trusted WLAN access and establishes layer-2 connection with the TWAP using the 802.11 WLAN association procedure.

(2) The N5CW device initiates the EAP-AKA' procedure by providing its NAI. The TWAP uses the realm part of the NAI to select the TWIF and sends an AAA request to it. If this is the first time this N5CW registers to this PLMN, then the NAI contains this device's SUCI. In this case, the SIDF de-conceals the SUCI to obtain SUPI. If the N5CW has previously registered to the 5GC over 3GPP access, then the NAI contains the 5G-GUTI assigned to it at that time. In this case, TWIF selects an AMF from 5G-GUTI.

Figure 2.14 Trusted WLAN access (N5CW) authentication procedure.

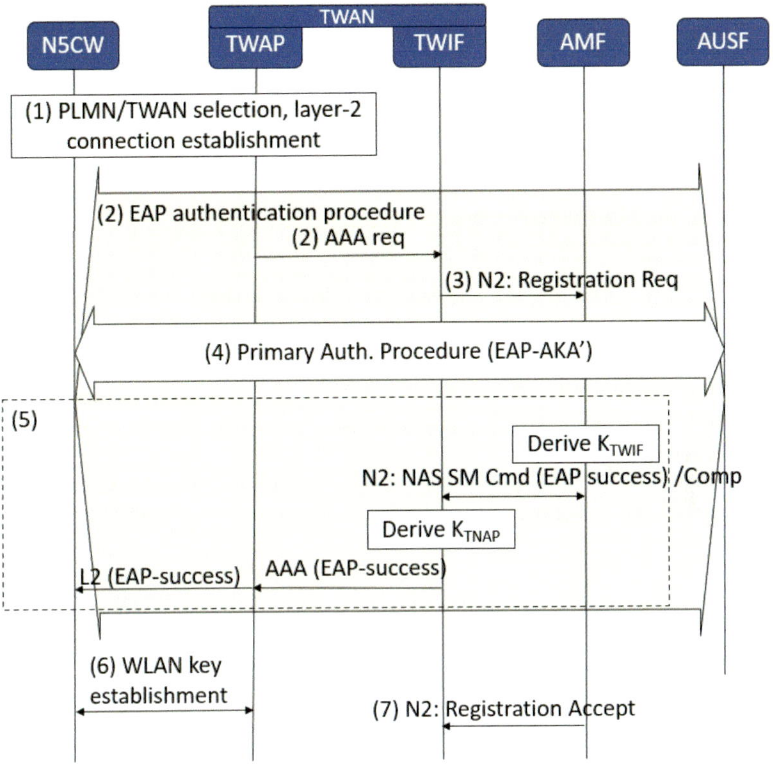

(3) The TWIF sends an NAS *registration request* message to the AMF on behalf of the N5CW. At this time, the TWIF uses default values in this message common to all N5CW devices.

(4) The AMF triggers the EAP-AKA' procedure, as discussed in Section 2.0, between the N5CW and the AUSF. Because the N5CW device does not support NAS signaling, EAP-AKA' messages between the TWIF and the 5GC (AMF and AUSF) are encapsulated in NAS authentication messages and terminate at the TWIF. Between the N5CW device and the TWIF, EAP-AKA' messages are encapsulated within the AAA (to the TWAP) and in the IEEE 802.11/1x layer-2 frame (to the N5CW).

(5) After successful completion of the EAP-AKA' procedure, the AMF derives K_{TWIF}. An NAS security context between the TWIF and the AMF contains null encryption and null integrity protection. A part of the reason why no NAS security context is established is because the N5CW does not support NAS signaling, thus the NAS counter is not relevant to the N5CW. This is similar to establishing an emergency call.

The AMF sends an NAS security mode command (SMC) which contains EAP success to the TWIF. The TWIF returns an NAS *security mode complete* message to the AMF. The AMF sends an N2 *initial context setup request* message containing the derived K_{TWIF}. From K_{TWIF}, the TWIF derives K_{TNAP} and forwards it, along with an *EAP success* message, to the TWAP which forwards them to the N5CW device.

(6) Using the K_{TNAP} key as the PMK (as in trusted non-3GPP access in Section 2.0), the N5CW and TWAP (WLAN AP) execute a four-way handshake to establish the security context between them to protect traffic over the air.

(7) The AMF sends the NAS *registration accept* message to the TWIF.

2.8 Wireless and Wireline Access

The wireless and wireline convergence for the 5G system is enabled due to the introduction of 5G-RG and FN-RG. This architecture is shown in Figure 2.15, where 5G-capable UEs can connect to 5GC through an RG via a wireline access network (W-5GAN). This scenario supports authentication of authenticable non-3GPP devices (AUN3) – devices that do not possess 5G credentials.

In the remainder of this section, we discuss residential gateway scenarios (5G-RG and FN-RG), followed by a no-5G credential device (AUN3).

Figure 2.15 Wireless and wireline convergence overview.

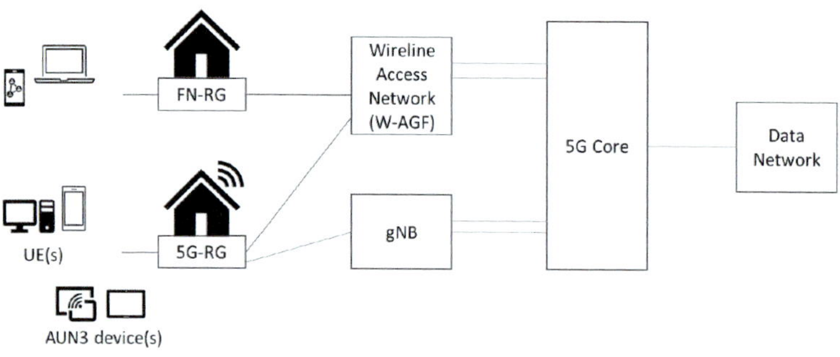

2.8.1 Residential GW (5G-RG, FN-RG)

There are two types of residential gateways: 5G-RGs and FN-RGs. We discuss how they are connected to the 5GC in detail.

5G-RG: A 5G-RG is equivalent to a normal UE and can connect to 5GC via wireline access network (W-5GAN) or via fixed wireless access (FWA)/NG RAN or via both accesses. The 5G-RG also acts as an end point of the N1 interface and provides the NAS signaling connection to the 5GC on behalf of the AUN3 devices behind the 5G-RG. In the case of 5G-RG connecting to 5GC via W-5GAN, i.e., 5G-RG establishes a wireline access control plane protocol (W-CP) connection with a W-5GAN, W-CP protocol stack messages are used between the 5G-RG and the W-5GAN for encapsulating NAS messages. The 5G-RG sends a message using W-CP protocol stack that can contain a *registration request* message containing SUCI/5G-GUTI. 5G-GUTI is used in the case where the 5G-RG has registered to the same AMF through the NG RAN and if this is the first time that the 5G-RG connects to the 5GC through W-5GAN. The wireline access gateway function (W-AGF) selects an AMF and forwards the *registration request* message to it. The AMF decides to authenticate the 5G-RG if a SUCI is received or perform NAS SMC based on the 5G-GUTI by reusing the existing security context. In case the AMF initiates the UE authentication, the authentication specific message exchange with the AUSF and the UDM can be the same as normal UEs and a suitable authentication method can be selected and run with the 5G-RG. Between the AMF and the UE (i.e., 5G-RG), the authentication messages are encapsulated within NAS authentication messages and these messages are carried in the N2 signaling between the AMF and the W-AGF, and then are further encapsulated using W-CP protocol stack message.

Following successful authentication, K_{AUSF}, K_{SEAF}, and K_{AMF} key derivation occurs as normal UEs. The NAS COUNTs associated with a NAS connection identifier “0x02” are set at the 5G-RG and the AMF. The AMF sends an SMC message to the UE (5G-RG) in order to activate a NAS security associated with the NAS connection identifier “0x02”. This message is first sent to W-AGF within an N2 message. If the EAP-AKA’ method is used for authentication, the AMF encapsulates the *EAP-success* message received from the AUSF within the SMC message. Upon performing a successful NAS SMC, the AMF sends an N2 *initial context setup request* message to the W-AGF which includes the $K_{WAGF.}$ The key K_{WAGF} may be used by the 5G-RG and W-AGF to secure the connection between them. The AMF sends an *NAS registration accept* message and the W-AGF forwards it to the 5G-RG over the established W-CP connection. All further NAS messages between the UE and the W-AGF will be sent over this established W-CP connection.

FN-RG: An FN-RG can connect to the 5GC via a wireline access network (W-5GAN). The W-AGF performs the registration procedure on behalf of the FN-RG and the W-AGF acts as the end point of the N1 interface and provides the NAS signaling connection to the 5GC on behalf of the FN-RG. Since the FN-RG is a non-wireless entity defined by BBF or CableLabs, it does not support the N1 interface. The W-AGF provides the N1 connectivity on behalf of the FN-RG. The W-AGF may authenticate the FN-RG; this is controlled by local policies.

It is assumed that there is a trust relationship between the wireline operator that manages the W-5GAN and the PLMN operator that manages the 5GC. The AMF trusts the W-5GAN based on mutual authentication executed when security is established on the interface between the two using NDS/IP or DTLS. The FN-RG is authenticated by the W-AGF. The authentication method used for FN-RG is defined by BBF or CableLabs, thus out of scope of 3GPP.

The W-AGF performs initial registration on behalf of the FN-RG. The W-AGF generates a *registration request* message and sends it to the AMF over the N2 interface. This *registration request* message contains the SUCI of the FN-RG. The N2 message contains an indication that the W-AGF has authenticated the FN-RG. The AMF selects an AUSF and sends *authentication request* message along with the SUCI of the FN-RG. It also contains the authenticated indication generated by the W-AGF. The AUSF forwards the SUCI and the authenticated indication to UDM in an *authentication get request* message to the UDM. The UDM determines, after de-concealing the SUCI to the SUPI, based on the subscription profile of the SUPI and the authenticated indication that authentication has been completed by the W-5GAN, that authentication by the home network is not required for the FN-RG. The UDM sends an *authentication get response* message to the AUSF with the SUPI and an indication that authentication by the home network is not required. After checking the UDM's indication, the AUSF does not perform authentication and sends an *authentication response* message to the AMF with SUPI and the indication that authentication by the home network is not required (i.e., no K_{SEAF} is included). After checking the indication, the AMF establishes the NAS security between the AMF and W-AGF with NULL encryption and NULL integrity protection. Further the AMF sends a *registration accept* message to the W-AGF which contains 5G-GUTI and other parameters. The W-AGF then sends a *registration complete* message back to the AMF and the W-AGF stores the 5G-GUTI for use in later NAS procedures.

2.8.2 No-5G credential device (AUN3)

To support an authenticable non-3GPP device (AUN3), the 5G-RG also acts as an end point of the N1 interface and provides the NAS signaling connection to the 5GC on behalf of the AUN3 devices. In this case, AUN3 devices can be registered and authenticated by the 5GC using EAP-AKA' method. There are two AUN3 device categories depending on whether the device's capability to support 5G key hierarchy or not.

(1) AUN3 devices that do not support 5G key hierarchy: The 5G-RG sends an NAS *registration request* message to the AMF, including the SUCI of the AUN3 device (received from the AUN3

device during *EAP ID request/response* messages) and an AUN3 device indicator. The AMF initiates primary authentication via the AUSF and the UDM. In this case, the UDM selects an EAP-AKA' based on de-concealed SUPI and the received AUN3 device indicator. According to the AUN3 subscription data along with the authentication vector, the UDM sends the MSK indicator to the AUSF to indicate that the AUN3 device does not support the 5G key hierarchy. Further EAP-AKA' method-specific *challenge request/response* messages are exchanged. If successful, based on the MSK indicator, the AUSF generates the MSK according to RFC 5448 [5], and the AUSF does not generate the K_{AUSF}.

Further the AUSF sends an *authentication response* message to the AMF/SEAF including the *EAP-success*, the MSK, and the SUPI. Based on the received MSK, the AMF does not generate the K_{AMF}. Instead, the AMF sends the *EAP-success* message and the MSK to the 5G-RG in N1 message/*NAS security mode command/authentication result* messages. If it is a NAS *security mode command*, the AMF uses NULL encryption and NULL integrity protection since the NAS security context is not applicable to AUN3 devices. The 5G-RG sends the *EAP-success* message to the AUN3 device in a layer-2 frame. If the layer-2 connection is over WLAN (IEEE 802.11), the AUN3 device and the 5G-RG use the first 256-bit of the MSK as the PMK, from which the WLAN keys are derived. The AUN3 and the 5G-RG perform four-way handshaking to establish WLAN secure connection.

(2) AUN3 devices that support 5G key hierarchy: The initiation of registration and authentication is similar to the AUN3 devices that do not support 5G key hierarchy. In this case also, the UDM selects the EAP-AKA' method based on the de-concealed SUPI and the received AUN3 device indicator. Further, EAP-AKA' method-specific *challenge request/response* messages are exchanged. If successful, a 5G key hierarchy is derived for the device; the AUSF generates and provides K_{SEAF} (i.e., an anchor key) to the AMF/SEAF along with the *EAP-success* message and the SUPI. In the final *authentication* message from the home network, the SEAF derives the K_{AMF} from the K_{SEAF} and send it to the AMF. The AMF further derives a K_{WAGF} key and sends an *EAP-success* message and K_{WAGF} to the 5G-RG in N1 *NAS security mode command/authentication result* messages. If it is a *NAS security mode command* message, the AMF uses NULL encryption and NULL integrity protection since the NAS security context is not required in this scenario. The key K_{WAGF} may be used by the 5G-RG and W-AGF for securing the connection between them. The 5G-RG sends the *EAP-success* message to the AUN3 device in a layer-2 frame. If the layer-2 connection is over WLAN (IEEE 802.11), the AUN3 device and the 5G-RG use K_{WAGF} as the PMK, from which the WLAN keys are derived. The AUN3 and the 5G-RG performs four-way handshaking to establish a secure WLAN connection between them.

2.9 Non-public Network (NPN) Authentication

In this section, we discuss the feature that supports non-public networks (NPNs) in a 5GS. Most of the security procedures are the same as public networks. Therefore, we focus on the minor exceptions with respect to the type of NPN, i.e., PNI NPN and SNPN respectively.

2.9.1 Public network integrated NPN (PNI-NPN)

A public network integrated NPN (PNI-NPN) is a type of NPN that is deployed with the support of a PLMN. PNI-NPN may be deployed as a private network slice (i.e., a dedicated network slice by the PLMN for the sole use of the owner), or it may consist of a combination of the PLMN's network and the NPN owner's network. For PNI-NPN, the primary authentication is performed by the public network as described in Section 2.2. After successful primary authentication, further secondary authentication as described in Section 2.3 and slice-specific authentication as described in Section 2.4 can take place.

2.9.2 Standalone NPN (SNPN)

A standalone non-public network (SNPN) is a type of a private 5G network that is separately deployed from public 5G networks. As such, a private 5G network is standalone and independent from public 5G networks. One of the key characteristics of an SNPN is that authentication methods other than AKA may be used. When an AKA-based authentication method is used, the primary authentication procedure described in Section 2.2 applies. When an authentication method other than 5G AKA or EAP-AKA' is used, only the non-AKA specific parts of Section 2.2 apply. An example of UE authentication in SNPN is the EAP-TLS method.

In SNPN, the authentication server can be either internal or external; the internal authentication server is the AUSF, and the authentication method can be 5G-AKA or EAP-AKA´ as described in Section 2.2 or can be EAP-TLS. The UDM determines whether an internal or external authentication server is used for primary authentication. The external authentication server can be classified as one of the two following cases:

(1) When the external authentication server is AAA: The primary authentication procedure using AAA as the authentication server is described in TS 33.501 Annex I.2.2.2 [3]. This case allows UEs to access an SNPN that makes use of a credential management system managed by a credential provider external to the SNPN. In this case, the authentication server role is assumed by the AAA server. The AUSF takes the role of the EAP authenticator and interacts with the AAA server to execute the primary authentication procedure.

(2) When the external authentication server is an AUSF: The primary authentication procedure using an AUSF and UDM is described in TS 33.501 Annex I.2.4 [3]. The SNPN uses AUSF and UDM as credential holders for primary authentication and authorization. It enables UEs to access an SNPN that makes use of a credential management system managed by an AUSF and a UDM internal to the SNPN.

2.9.3 Non-seamless WLAN offload (NSWO)

Non-seamless WLAN offload (NSWO) is an optional capability in UEs that support WLAN radio access. A UE supporting NSWO may, while connected to WLAN access, route specific IP flows via the WLAN access without traversing the 3GPP core network. 5G NSWO uses the EAP-AKA' method for authentication, as defined in RFC 5448 [5]. The EAP-AKA' implementations are expected to comply with the EAP-AKA' profile specified in TS 33.501 Annex F [3]. A new network function, called the NSWO function (NSWOF), supports authentication for NSWO in the 5GS. The NSWOF interacts with the WLAN access network (AN) over an SWa interface and communicates with the AUSF using a service based interface (SBI). NSWOF acts as an SBI/AAA proxy between the AUSF and the WLAN AN.

An HPLMN that supports 5G NWSO configures the UE to use 5G NSWO. This configuration can be stored either in the USIM or in the ME. If both of them are configured, the configuration in the USIM takes precedence. If a UE supports 5G NSWO and it is configured to use it, the UE always uses this feature. Otherwise, the UE uses EPS NSWO. Such a configuration ensures that the UE supporting 5G NSWO cannot be downgraded to use EPS NSWO.

The network may support both 5G NSWO and EPS NSWO. In this case, the routing of the AAA messages is determined by the network based on the realm part of the UE identity (e.g., *epc.mnc<MNC>.mcc<MCC>.3gppnetwork.org* for EPS NSWO, or *5gc-nswo.mnc<MNC>.mcc<MCC>.3gppnetwork.org* for 5G NSWO). Which entities in the network perform this routing decision depends on the network configuration.

The UE establishes a WLAN connection between the UE and the WLAN AN, using procedures specified in IEEE 802.11. The WLAN AN sends an *EAP identity/request* message to the UE and it receives an *EAP response/identity* message from the UE. If the UE determines to use the NSWO service, the UE use the SUCI in NAI format as its identity irrespective of whether SUPI type configured on the USIM is IMSI or NAI. In the former case, the UE constructs the SUCI in NAI format with a username containing the encrypted MSIN and the realm part containing the MCC/MNC.

The *EAP response/identity* message is sent over the SWa interface to the NSWOF which is identified by the realm part of the SUCI. The NSWOF sends the *authentication request* message to the AUSF including SUCI, AN identity and NSWO indicator. The NSWO indicator is used to indicate to the AUSF that the authentication request is for non-seamless WLAN offload purpose. The NSWOF set *"5G:NSWO"* as the AN identity value. Based on this NSWO_indicator, the AUSF, acting as the EAP authentication server, sends an *authentication get request* message to the UDM, including SUCI and the AN identity and NSWO indicator. The UDM after de-concealing the SUCI to a SUPI, based on the NSWO indicator, the UDM/ARPF selects the EAP-AKA' authentication method and generates an authentication vector using the AN identity as the KDF input parameter. In the *authentication_get response* message to the AUSF, the UDM includes the EAP-AKA' authentication vector (i.e., a tuple of RAND, AUTN, XRES, CK′ and IK′) and may include the SUPI.

The EAP-AKA' specific *authentication challenge* message exchange takes place. The NSWOF sends the *EAP-request/AKA'-challenge* message to the WLAN AN over the SWa interface. The WLAN AN forwards this message to the UE. Further verifications and key derivation take place similar to the EAP-AKA' method. In the NSWO authentication, the UE does not generate a K_{AUSF}. The UE sends the *EAP-response/AKA'-challenge* message to the WLAN AN. The WLAN AN forwards this message over the SWa interface to the NSWOF. The NSWOF sends the *authenticate request* forwarding the *EAP-response/AKA'-challenge* message to the AUSF. The AUSF verifies the received response (RES) against the stored and expected response (XRES). Upon successful verification, the AUSF proceeds to the next step; otherwise it returns an error to the NSWOF.

Based on the NSWO indicator received, the AUSF derives the required MSK key from CK' and IK' and as described in RFC 5448 [5]. The AUSF does not generate the K_{AUSF}. Then the AUSF sends the *authentication response* message with *EAP-success* and the MSK key to the NSWOF. The AUSF optionally provides the SUPI to the NSWOF. The NSWOF sends the *EAP-success* message and the MSK to the WLAN AN over the SWa interface. The *EAP-success* message is forwarded from the WLAN AN to the UE. Upon receiving the *EAP-success* message, the UE derives the MSK, if it has not yet done so. The UE uses the first 256-bit of the MSK as the PMK to perform a four-way handshake to establish a secure connection with the WLAN AN.

In the roaming scenario, the HPLMN may have a roaming agreement with a VPLMN for NSWO roaming. A roaming UE configured by the HPLMN to use 5G NSWO may try to register to a WLAN AN that may advertise the UE's HPLMN or a VPLMN (with which the HPLMN has a roaming agreement for NSWO roaming).

2.10 UAV USS Authentication and Authorization (UUAA)

The UAV USS authentication and authorization (UUAA) procedure [14] enables UAV authentication and authorization by a UAS service supplier (USS) before the connectivity for UAS services is enabled by the 3GPP network (i.e., 5GS or EPS) as shown in Figure 2.16.

Figure 2.16 Overview of UAV authentication for UAS services in 5GS and EPS.

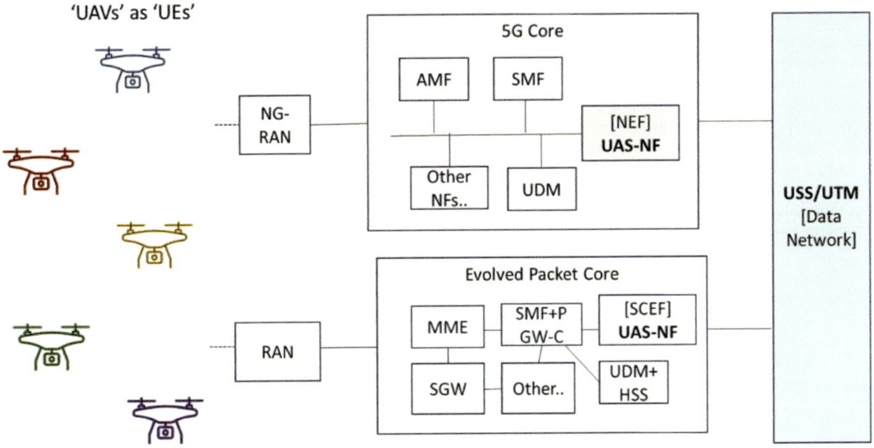

A UAV performs UUAA with the USS/UAS traffic management (USS/UTM) only after the UAV (UE) has successfully completed primary authentication. UUAA may be triggered under several different conditions: (1) the AMF triggers it when the UAV is registered with 5GS, or (2) the SMF triggers it during the PDU session establishment procedure as shown in Figure 2.17, or (3) the USS triggers it for re-authentication if the USS has authenticated the UAV. UUAA is performed between the UAV and the USS. The UAV is authenticated based on the CAA-Level UAV ID and credentials associated with the CAA-Level UAV ID. The *authentication* messages are included in a transparent container and conveyed between the UAV and the USS via the 3GPP UAS NF. On successful completion of UUAA, the USS can send UAS security information in the UUAA *authorization payload* to the UAV as shown in the following steps.

There are two cases: (A) UUAA during the UE registration procedure, and (B) UUAA during the UE PDU session establishment procedure. We describe the steps separately below.

Case A: UUAA during the UE registration procedure:

(1) The UE sends a *registration request* message to the AMF. The UE may provide a CAA-level UAV ID, and optionally an USS address/IP address, to indicate the request is registering for UAS services. The AMF completes NAS security set up including primary authentication as needed. Then the AMF returns a *registration accept* message to the UE and indicates that UUAA is pending. The UE may send a *registration complete* message to acknowledge the AMF.

Figure 2.17 UAV USS authentication and authorization (UUAA) procedure over 5GS during registration or PDU session establishment.

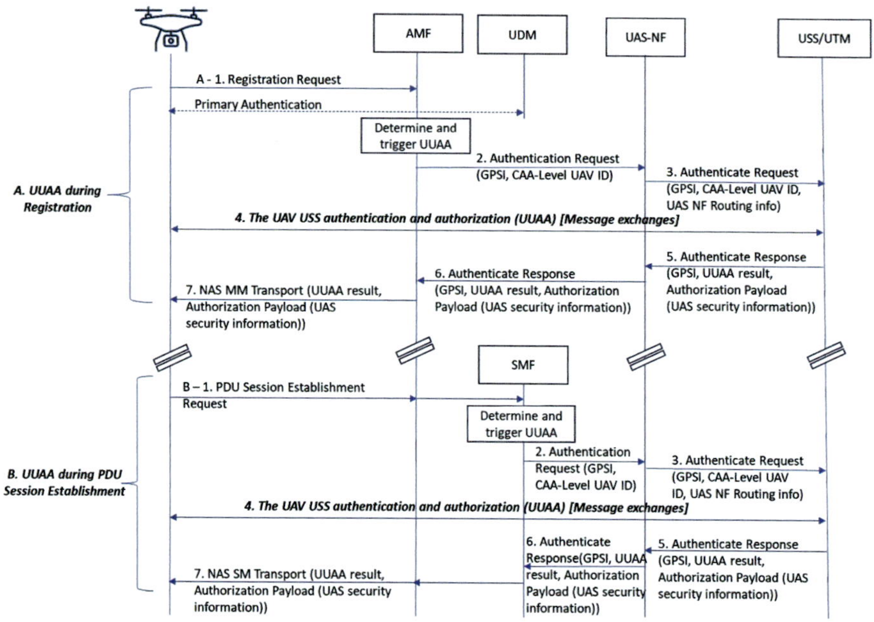

(2) After the successful primary authentication, the AMF determines if UUAA is required for the UE. The AMF triggers the UUAA procedure if the UE has provided a CAA-level UAV ID and has a valid aerial UE subscription in the UDM. The AMF may skip UUAA if the UE has previously completed the UUAA procedure successfully and the UE UUAA is current, i.e., the UE's authentication and authorization has not been revoked after the last successful UUAA.

The AMF sends the *authentication request* message to the UAS-NF. The UAS network function is supported by the NEF (in the case of the 5GS) or SCEF+NEF (in the case of the EPS) and is used for external exposure of services to the USS.

(3) The UAS NF resolves the USS address using the CAA-level UAV ID or uses the provided USS address. Only an authorized USS is used in order to ensure only legitimate entities can provide authorization for UAVs. The UAS NF sends an *authentication request* message to the USS. This message includes the GPSI, the CAA-level UAV ID, and a UAS NF routing information (e.g., an FQDN or IP address) which uniquely identifies the UAS NF located in the 3GPP network that handles the UAV related messages exchanges with the corresponding external USS/UTM and the transparent container.

(4) The USS and the UE exchange authentication messages. Multiple round-trip messages may be needed as required by the authentication method used by the USS.

(5) The USS replies to the UAS NF by sending the *authentication response* message. This message includes the GPSI, the UUAA result (success or failure), the authorized CAA-level UAV ID, and a UUAA authorization payload that contains UAS security information.

(6) The UAS NF sends an *authentication response* message to the AMF, including the GPSI, the UUAA result (success or failure), the authorized CAA-level UAV ID, and the UUAA authorization payload.

(7) The AMF sends the UUAA result (success or failure) to the UE in an NAS *MM transport* message. Further, the AMF stores the results, together with the GPSI and the CAA-level UAV ID.

Case B: UUAA during the UE PDU session establishment procedure:
If the SMF determines to trigger UUAA, the same procedure described in case A, relative to UUAA, can be executed during PDU session establishment:

(1) The UE sends a *PDU session establishment request* message to the SMF including a CAA-level UAV ID to indicate the request is for UAS services.

The SMF determines if UUAA is required for the UE. UUAA is only triggered if the UE has provided a CAA-level UAV ID and has a valid aerial UE subscription. The SMF may skip the UUAA procedure, if a successful UUAA result is available from the last successful PDU session establishment with the same USS/DN.

(2–6) The SMF triggers the UUAA procedure if determined as necessary. In this case, the same procedure as described in steps 2–6 in case A is performed.

(7) The SMF sends the UUAA result (success/failure) to the UE in the NAS *SM transport* message. Further, the SMF stores the result, together with the GPSI and the CAA-level UAV ID values.

2.11 Summary and Conclusion

This chapter discussed an overview of different authentication procedures used for various purposes in the 5G system. Primary authentication is used for the mutual authentication of the UE and the network with its new enhancements such as enhanced home control and the possibility to trigger the primary authentication over non-3GPP accesses (e.g., WLAN), resulting in multiple NAS connections. The primary authentication of a UE can be triggered by either the home network or by the serving network based on operator's authentication policies. Non-3GPP access to the 5G network can include trusted non-3GPP access networks and untrusted non-3GPP access networks which are IP-based access networks. Further, wireline access to the 5G core network can also be facilitated by authentication procedures for residentials gateways such as 5G-RG and FN-RG for UEs located behind these RGs, and for N5CW devices and AUN3 devices that either support or do not support 5G key hierarchy. Whether a non-3GPP access network is trusted or untrusted is determined outside the access network's characteristics. In a non-roaming scenario, it is the home network operator's policy whether a non-3GPP access network is used as trusted or untrusted non-3GPP access. When the home network considers one or more of the security feature groups (and other aspects) provided by the non-3GPP access network not sufficiently secure, the home network may identify the non-3GPP access as an untrusted non-3GPP access. Primary authentication uses either EAP-AKA' or 5G-AKA method for public networks and mandates the subscription security credentials and shared secret to be stored on the USIM in the UE and the UDM/UDR of the operator network. The primary authentication can also be based on any of the optional EAP-based authentication method(s). Following successful primary authentication, the UE is allowed to connect with the network and the connection is secured using the derived session keys. Secondary authentication is an optional procedure, and it can use any EAP method outside of the registration procedure. It may be required when the UE is requesting a PDU session towards an external DN. This is either based on subscription feature or network configuration in the SMF, which takes the role of the EAP authenticator. The signaling messages towards the SMF are encapsulated in an NAS container and then sent as user plane traffic from the SMF via the UPF to the AAA server.

Slice authentication is required to access certain S-NSSAIs. It is performed together with the registration procedure and requires successful primary authentication as a prerequisite before accessing the AAA server for the EAP-based slice-specific authentication. The EAP authenticator role is performed by the AMF instead of the SMF and the authentication signaling traverses through the NSSAAF in the home network before reaching the AAA server and is not sent via user plane as in the secondary authentication.

AKMA is designed over the primary authentication and enables UE applications related to 3GPP services (e.g., additional network services) and 3rd party services to utilize the 3GPP credentials (e.g., K_{AUSF}) for establishing application security (e.g., K_{AKMA} and K_{AF}) so that the UE can exchange data with an application server in a secure manner.

Moreover, UAVs with aerial subscription with the mobile network operator act as UEs. A UAV is authenticated and authorized by the USS via the UUAA procedure with the support of the 3GPP system before the UAS services is enabled.

2.11.1 Outlook

Extensive work has been carried out on authentication methods involving different dedicated procedures to cater to various service needs, starting from

Release 15 (5G system) to the current Release 18 (5G-Advanced system). For the future Release 19, various security and service optimization topics are being discussed such as security for mobility over non-3GPP access scenarios to avoid full primary authentication in TR 33.702 [11]. Some of the scenarios covered in this study include: (1) UE connecting to the new TNAP within the same TNG, (2) AUN3 device connecting to a new 5G-RG under the same W-AGF, (3) N5CW device connecting to a new TWAP under the same TWIF, and (4) UE connecting to a new WLAN AP connected via the same NSWOF. Furthermore, in another study related to study on security for PLMN hosting an NPN (TR 33.757 [12]), SUPI privacy issues during primary authentication in case of PLMN hosting NPN scenarios are being studied, as the 5G system should enable a PLMN to host an NPN without compromising the security of that PLMN. In this scenario, the study investigates supporting necessary security mechanism in the 5G system to ensure the protection of the sensitive parameters against the risk caused by PLMN hosting NPN and vice versa. These ongoing studies ensure that 5G system security evolves along with the most recent architecture changes as needed by the service enhancements.

Acknowledgements

This work was supported in part by CyberSecurity Research Flanders with reference number VR20192203.

References

[1] 3GPP, TS 23.501 (v18.4.0), System architecture for the 5G System (5GS), Dec. 2023

[2] 3GPP TS 23.502 (v18.4.0), Procedures for the 5G System (5GS), Dec. 2023

[3] 3GPP TS 33.501 (v18.4.0), Security architecture and procedures for 5G System, Jan. 2024

[4] 3GPP TS 33.102 (v17.0.0), 3G security; Security architecture, Mar. 2022

[5] IETF RFC 5448, Improved Extensible Authentication Protocol Method for 3rd Generation Authentication and Key Agreement (EAP-AKA'), May 2009

[6] IETF RFC 3748, Extensible Authentication Protocol (EAP), June 2004

[7] 3GPP TS 33.220 (v18.2.0), Generic Authentication Architecture (GAA); Generic Bootstrapping Architecture (GBA), Dec. 2023

[8] 3GPP TS 33.163 (v17.0.0), Battery Efficient Security for very low Throughput Machine Type Communication (MTC) device (BEST), Dec. 2021

[9] 3GPP TR 33.835 (v16.1.0), Study on authentication and key management for applications based on 3GPP credential in 5G, July 2020

[10] 3GPP TS 33.535 (v18.2.0), Authentication and key management for applications based on 3GPP credentials in the 5G System (5GS), Dec. 2023

[11] 3GPP TR 33.702 (V0.1.0), Study on Security for mobility over non-3GPP access to avoid full primary authentication, March 2024.

[12] 3GPP TR 33.757 (V0.1.0), Study on security for PLMN hosting a NPN, March 2024.

[13] 3GPP TR 33.741 (v18.0.1), Study on home network triggered primary authentication, Apr. 2023

[14] 3GPP TR 33.256 (v18.1.0), Security aspects of Uncrewed Aerial Systems (UAS), Dec. 2023

CHAPTER

3

Nuances of 5G Security: Why not all 5G Deployments are Created Equal

Hans C. Rudolph

Deloitte Tohmatsu Cyber, Japan

Email: hans@hcrudolph.com

Abstract

5G has been praised by many as a step forward in terms of security and data protection. While it is true that it includes several new security and privacy features, it is also important to note that not everything contained in the technical specifications actually finds its way into real-world deployments. Local regulation, cost factors, and other operational considerations can result in operators deciding against rolling-out a given feature. The following text explores different 5G security controls and parameters, discusses to what extent network operators can determine their implementation, and the possible security implications of these decision.

3.1 Introduction

3.1.1 The success and limitations of global telecom standards

The field of telecommunications is a prime example of the importance and the success of international standards. Over the last 25 years, the mobile

DOI: 10.1201/9788743810902-3

industry – and specifically the 3rd Generation Partnership Project (3GPP)– has delivered comprehensive specifications for three generations of cellular communication systems. Without them, basic use cases that are ubiquitous today, such as using our mobile phones abroad, would not be possible. These technical specifications, and the international standards based on them, are what makes the fundamental components and procedures globally interoperable.

However, in their broad scope and applicability also lies an inherent limitation of these standards: they usually do not go into great detail, and many things are either optional or left entirely up to those implementing them. In the context of 5G security, that generally means anything concerning communication within a single network operator's security domain can hardly be mandated. Moreover, geopolitical, and regulatory factors rule out one unanimously agreed definition of "5G security". Hence, 3GPP specifications always contain a certain degree of optionality. This is where mobile operators have to take informed decision on the security baseline of their services and enforce network-wide policies accordingly.

This text highlights some of the optional 3GPP features with regards to 5G security and explores how network operators can utilize them to ensure subscribers and their data are protected in the best way possible. It also outlines the purpose and limitations of these features in an attempt to dispel some misunderstandings about standard 5G security controls. Of course, the following is not an exhaustive list of optional features in the 5G security specifications. Instead, the focus lies on controls that directly contribute to protecting subscribers and subscriber data, rather than the 5G infrastructure itself.

3.1.2 Basic concepts and terminology

Before looking into specific 5G security controls and parameters, it is worth defining some basic terminology first. These terms will be used accordingly throughout the remainder of this text.

Cellular mobile networks are generally comprised of three main components: the user equipment (UE), the radio access network (RAN), and the core network. The user equipment is the mobile terminal that subscribers use to connect to the mobile network (e.g., a mobile phone). The RAN is largely comprised of radio transceiver stations responsible for providing wireless coverage within the service area of the network. As such, it transforms digital signals from the network into radiofrequency waves that are sent to the user equipment and vice versa. The core network contains most of the "business logic" required to provide mobile communication services. For example, it ensures that subscribers are properly authenticated before accessing services, keeps track of their usage so they can later be billed, and forwards their traffic to and from external data networks, such as the public internet.

The 5G specification defines a number of deployment options for the 5G system, intended to make the transition from existing deployments of earlier generations – particularly 4G LTE – easier for network operators. At a high level, one can distinguish 5G standalone (SA) and 5G non-standalone (NSA). The former is comprised of 5G technology components only, both in the RAN and in the core network. The latter describes mixed deployments which partially utilize 4G LTE components. While there are several types of NSA deployments which differ in various technical details, one of the more popular options is one that uses a 5G RAN and a 4G LTE core network. Although subscribers can take advantage of the increased throughput of 5G in this case, there is little improvement in terms of security: since many 5G security features are provided by the core network, they are not available to subscribers unless they are connected to a 5G SA deployment.

Lastly, there is a distinction to be made between the different parties involved in providing end-to-end mobile services. Generally, one can distinguish between the home network operator and the serving network operator. The home network operator is the one that the subscriber has its service contract with. It is also the entity that manages the mobile subscription, the authentication credentials, and writes a bill at the end of the month. In addition, there are serving network operators which operate the mobile network that the subscriber is connected to at a given point in time. This can be the same entity as the home network operator or a different one, depending on the subscriber's current location. For example, when a subscriber is roaming abroad, the chances are that they connect to a mobile network operated by a different entity than their home network operator.

The following section explores optional 5G security controls and parameters categorized by the parties responsible for implementing or controlling them. Section 3.2.1 describes security controls under control of the home network operator, while Section 3.2.2 focuses on serving network operator-controlled aspects. Section 3.2.3 looks at security controls jointly provided by the home network and serving network operator. Lastly, Section 3.2.4 describes security controls that are individually determined by home network and serving network operator for their respective network environments.

3.2 Optional 5G Security controls

3.2.1 Security controls under control of the home network operator

The first category comprises security controls that are solely controlled by the home network operator. Since this is the entity managing the users' subscription, these controls are mainly related to the subscriber's permanent identity and the primary authentication procedure.

Before any mobile communication can happen, the home network operator needs to create a subscriber profile. This profile is a data set containing basic information about a user's subscription, for example, which services they are allowed to access. Part of the subscriber's profile is the long-term key, which is used for mutual authentication between user equipment (UE) and home network and for deriving the session keys protecting the communication over the air interface between the UE and the serving network. An important decision to be made by the home network operator is the length of this long-term key. While historically 128-bit keys were used, the current 3GPP technical specifications also support 256 bits. This is in line with the rest of the 5G system that has been designed to support 256-bit cryptographic keys from the start. However, as of March 2024 the technical specifications for the USIM still allow for the provisioning of 128-bit long-term keys. Of course, deriving 256-bit session keys from a root key that is only 128-bit long has no benefit in terms of security, as the entropy is limited by the initial 128 bits. Currently, the practical implications of this are minimal, because the security algorithms for protecting the air interface on NAS and AS layers only support 128-bit keys as well. So, for now, the associated session keys are simply truncated to 128 bits. However, 3GPP is already discussing the adoption of 256-bit cryptographic algorithms to be used on the air interface, and it is highly likely that these will be featured in an upcoming release. At least at this point, network operators should phase out shorter long-term keys and instead provision 256-bit keys to not unintentionally limiting the available entropy (see also [1], [4]).

A new security feature introduced with 5G, but also available to previous generations, is the long-term key update process (LTKUP). As the name suggests, this process allows the home network operator to provision a new long-term key to a subscriber's USIM as part of their existing subscription. This may be required for a number of reasons. The related 3GPP TR 33.935 [5] mentions the following scenarios, for example:

- Individual subscription – [long-term key] exposed
- Batch of subscriptions – [long-term key] exposed

- Loss of synchronization of long-term keys
- Undetected leakage of [long-term key].

Figure 3.1 LTKUP implementation options.

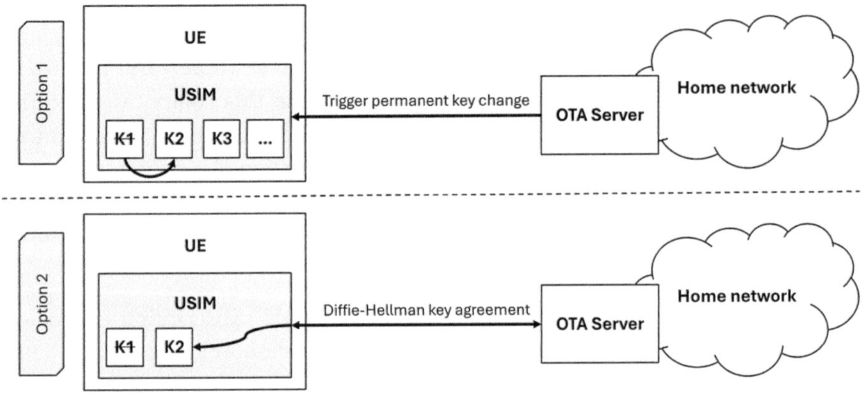

While LTKUP cannot prevent these incidents from happening, it provides the home network operators mechanisms for recovering from them. Specifically, 3GPP defines two viable solutions, both of which rely on over-the-air (OTA) communication between the home network operator and the UE. OTA involves the transmission of a specially crafted, protected SMS message carrying commands or data from an OTA server to the subscriber's UE which are passed on to the USIM (Figure 3.1).

The first LTKUP solution is the naïve approach of provisioning multiple sets of long-term keys into the USIM during the initial provisioning, i.e., the start of the subscription. This allows the home network operator to simply trigger the switch to a new set of keys via the appropriate OTA command. The second solution generates a new set of keys on-the-fly via the OTA channel. For this purpose, a Diffie–Hellman key agreement is performed between the home network and the USIM.

The LTKUP feature is entirely optional, and while it is not limited to 5G networks, it requires several prerequisites to be met. That is, the home network operator needs to have an OTA server as well as a subscriber database and USIM that support either of the LTKUP solutions outlined above (see also [5]).

5G is the first mobile generation that makes it possible to avoid transmitting the subscriber's permanent identity over the air in cleartext. As shown by security researchers in the past, leaking this information can put subscribers at risk of a plethora of targeted attacks and even location tracking and thus is to be prevented. The 5G system does so by using public-key cryptography to encrypt the subscription permanent identifier (SUPI) in the UE before it is sent over the air. However, it is important to note that this is an optional feature. For it to take effect, the home network operator needs to explicitly enable it in the subscription profile and provision the subscriber's UE with a network-specific public key for encryption which can only be reversed using the associated private key. The operator has the choice to perform this SUPI encryption either directly in the USIM or outsource this computation to the subscriber's mobile device. The first option may be of interest for operators that either wish to make use of non-standard encryption algorithms, or do not want to rely on the handset's implementation. The second option only allows operators to indicate the priority of standardized encryption algorithms to be used by the mobile device.

The result of the SUPI encryption is the so-called subscription concealed identifier (SUCI) which is then used for the primary authentication in lieu of the cleartext SUPI (Figure 3.2). In the home network operator's environment this concealed identifier is decrypted using the operator's private key to perform the authentication procedure.

Important to note is that, upon successful authentication, the home network also provides the SUPI to the serving network. Therefore, SUPI encryption only protects the subscriber's permanent identity from eavesdroppers on the air interface, but not from a potentially malicious serving network (see also [1]).

Figure 3.2 Use of SUCI and SUPI during primary authentication

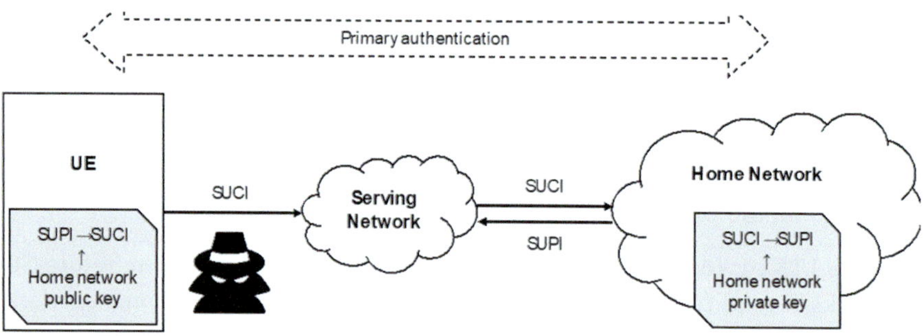

Another newly added security feature on steering of roaming (SoR) allows home network operators to provision a list of preferred serving networks to subscriber's UEs. This is relevant in roaming situations, when the UE has to select a foreign network to attach to. In most situations, it is in the interest of the subscriber to select a network that its home network operator has a preferential relationship, resulting in reduced roaming charges. In case a list of preferred serving networks has been provisioned, the selection will be made according to the priority defined in there. While the functionality itself is not entirely new, its implementation in 5G is different. Previously, delivery of such a list could be performed via OTA SMS, which could theoretically be blocked by malicious serving networks without the home network noticing. This new feature instead relies on the normal control plane communication of the 5G system between the home and serving network. Since the UE will confirm the receipt of the list to the home network, it is no longer possible for the serving network to silently block or even modify this communication. This works because all implementations are required to support the steering of roaming mechanism. However, it is up to the home network operator whether or not to actually use it (see also [1]).

Increased home control is another optional security feature relevant for the prevention of fraudulent authentication requests sent from the serving network to the home network. The specific challenge addressed by this mechanism are malicious serving networks that would claim to serve subscribers that are not actually present in their network, leading to fraudulent charges against the home network operator and, eventually, the subscriber. This problem is addressed in the 5G specifications as follows. Using one of the primary authentication methods in 5G – 5G AKA or EAP-AKA' – the home network obtains confirmation of each successful or unsuccessful primary authentication run and stores the result for future reference. This is the extent to which this feature is mandated in the technical specifications. However, simply obtaining this information does little to prevent the afore-mentioned fraudulent behavior. It also needs to be linked to subsequent interactions between the home network and serving network that may then be accepted or denied. This could include, for example, the rejection of a certain request by the serving network based on previous authentication results or the time at which they occurred. The 5G specifications provide informative guidance about these considerations, but they fall short at mandating any concrete actions. Therefore, it is up to the home network operator to determine the appropriate logic and request its technology suppliers to implement the same (see also [1]).

3.2.2 Security controls under control of the serving network operator

The second category of security feature concerns those under control of the serving network operator. As mentioned before, this entity may be the same as the home network operator, depending on whether or not the subscriber is currently roaming.

The serving network operator controls the radio access network, since that is the point to which mobile subscribers connect to the network. As such, it also determines the security of the air interface between UE and base station.

First, this includes the user plane security policy which specifies whether or not to protect user plane traffic, i.e., the actual data sent and received by subscribers, in contrast to the control plane required to manage the subscriber's connection. In this policy the need for confidentiality protection and integrity protection can be controlled independently. Common to both of these controls, however, is that applying them to user plane traffic is only optional.

Second, the serving network operator also determines the priority of cryptographic algorithms to be used on the air interface. The need for prioritization stems from the fact that the 5G system supports three distinct encryption and integrity protection algorithms, denoted NEA and NIA, plus the NULL algorithm:

- NEA0 / NIA0, NULL algorithm (no protection)
- 128-NEA1 / 128-NIA1, based on the SNOW 3G algorithm
- 128-NEA2 / 128-NIA2, based on the AES algorithm
- 128-NEA3 / 128-NIA3, based on the ZUC algorithm.

During session establishment, UE and network will negotiate the cryptographic algorithms to be used. The UE sends its security capabilities to the serving network, which in turn selects the matching algorithm with the highest priority based on its local configuration. As such, it is up to the serving network to disable algorithms that are not required or do not offer effective protection. While implementations should not allow use of the NULL algorithm for any use cases other than unauthenticated emergency calls, implementation flaws can occur and should still be tested for by the network operator (see also [1]).

Aside from the air interface, it is the responsibility of the serving network operator to provide protection for the communication links connecting individual base stations (i.e., on the midhaul) and between the RAN and the core network (i.e., on the backhaul). These links can be secured using either the IPsec or DTLS protocol. To achieve this, base stations must be provisioned with the appropriate cryptographic material, such as public key certificates. 3GPP TS 33.310 [3] describes a certificate enrollment mechanism for this purpose. However, enforcing security on these midhaul and backhaul interfaces is optional. While there is an argument that the risk is limited if the infrastructure is private, network-level protection is particularly important in shared infrastructure scenarios, for example, when parts of the RAN are virtualized in a local data center (see also [1]).

The serving network further assigns a temporary identifier to each subscriber. The so-called 5G globally unique temporary identity (5G-GUTI) is used to identify the mobile handset in lieu of the SUPI, so it is never required to send the permanent identity over the air. Of course, failing to regularly reassign a 5G-GUTI after a certain amount of time has lapsed defeats the purpose of a temporary identity and may lead to privacy issues for the subscriber. Therefore, the 5G specifications define certain events in which a new 5G-GUTI must be assigned to the UE by the serving network. These include the paging procedure, i.e., once the network has located the UE by broadcasting its temporary identity, and during periodic registration updates. The time between which a periodic registration update must be performed is signaled to the UE by the serving network. As such, it is up to the serving network operator to determine a suitably small timer value to ensure frequent reassignment of the 5G-GUTI. In addition to the scenarios mentioned above, it is left to implementations to allow for even more frequent updates of the temporary identifier (see also [1]).

3.2.3 Security controls to be agreed between the home network and the serving network operator

The 5G system also features optional security mechanisms that require support from both the home network and serving network operator. One such example is home network triggered reauthentication. In some scenarios, the home network may decide that a new primary authentication should be performed even if the subscriber is already registered, for example, when a subscriber is handed over from a legacy network to 5G. Performing the primary authentication procedure again creates more confidence in the subscriber's identity and refreshes the security context. Even though 5G only allows direct handovers from 4G LTE, this network may in turn be interworking with 3G or even 2G networks. Allowing subscribers to move between those security domains without reauthentication should be prevented (see also [1]).

Another example of a security control that has to be aligned between the home network and the serving network is the protection of the interconnect between the two domains (Figure 3.3). Since the home network manages the subscriber profile, it is involved in every primary authentication procedure, regardless of where the subscriber is connecting from. Beyond that, the subscriber's user plane traffic may also be forwarded over the interconnect to the home network before being routed to external data networks, depending on the roaming agreement between both operators. The 3GPP security specifications define two primary elements for protecting interconnect traffic:

- Security edge protection proxy (SEPP) for control plane traffic
- User plane function (UPF), which includes the inter-PLMN user plane security (IPUPS) functionality.

IPUPS mainly serves the purpose of protecting the mobile network itself, rather than the user plane traffic. Its purpose is to ensure that incoming traffic from other roaming partners is associated with a valid subscriber session and discarding malformed packages. Security of the user plane traffic should be provided on the network layer via IPsec between the user plane functions of the serving and home network operator. Its use is optional and up to the two roaming partners' discretion.

Figure 3.3 5G interconnect security.

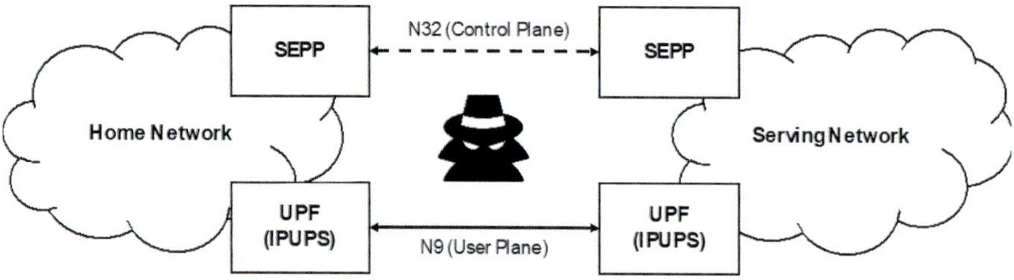

The security functionality of the SEPP is aimed at protecting both the network and the control plane traffic. On one hand, it is supposed to provide similar functionality as the IPUPS for filtering out malicious or malformed traffic by implementing mechanisms for anti-spoofing, rate limiting, and other filters based on message contents. On the other hand, it is also there to protect outgoing messages sent to other roaming partners. For this purpose, 3GPP specifies two mechanisms: end-to-end transport layer security (TLS) and application layer protection via the so-called PRotocol for N32 INterconnect Security (PRINS). This newly defined protocol offers roaming partners fine-grained control over what information to protect on the interconnect between them. While the 3GPP specification mandates a few crucial data points to be protected at any time, applying the same policy to the remaining data is not compulsory. This also includes the aforementioned permanent subscription identity (SUPI). Hence, operators aiming to achieve the maximum level of security and privacy for subscribers should mandate in their roaming agreements that as much information as possible is to be transferred in a protected manner. Beyond that, if operators do not necessarily depend on exchange providers to access the contents of the signaling data between them and their roaming partners (e.g. for the purposes of alleviating incompatibilities between them), there is also the option to use TLS on the interconnect. This should be the preferred way of integration since it offers end-to-end protection for complete communication (see also [1]).

3.2.4 Security controls under individual control of both the home network and the serving network operator

Lastly, some 5G security controls concern both the home network and the serving network operator equally. These include generic network security provisions that are recommended, but not mandated, in the 3GPP specifications. One example is the enforcement of mutual authentication as well as integrity and confidentiality protection between network elements, usually by means of IPsec or TLS. Some best practices for profiling these security protocols are outlined 3GPP TS 33.210 [2]. Ensuring that all network elements support the recommended cryptographic protocols and – equally important – disabling any other weak profiles is essential for ensuring effective cryptographic protection. As most network technology suppliers' solutions support a wide range of security algorithms for backwards compatibility reasons, making an informed decision on which algorithms to disable is vital. Of course, these security protocols also rely on some sort of cryptographic secret and thus require extra effort to be enabled. Historically, the argument against enforcing these controls uniformly across the network has often been that an operator's internal environment is considered "trusted" and hence security is provided implicitly. However, with 5G putting a strong emphasize on virtualization, distributed deployments, and edge computing, it becomes increasingly difficult to draw a line between "trusted" and "non-trusted" environments in modern network deployments. Moreover, with the adoption of API-driven communication and integration of third-party application functions in the 5G core, internal network components may be directly exposed to external systems. As such, the idea of implicit security is hardly a reasonable premise, and a shift to a zero trust architecture seems much more appropriate.

3.3 Considerations on Optionality

The naïve response to this list of optional security controls would be the question: Why not mandate all of them by default? Surely, it would be best to prescribe a single standard for strong security once and enforce that across mobile networks globally. It turns out that there are several reasons why this approach is not realistic, or even desirable.

Regulatory differences: National regulations for Telecom providers and critical infrastructure vary significantly across different countries and regions. Mandating specific security controls in the 3GPP technical specifications might conflict with local regulations or fail to meet the specific requirements of certain jurisdictions. Providing guidelines rather than mandates allows operators to comply with local laws while adhering to best practices.

Technological diversity and operator flexibility: The 5G standard intentionally allows for a wide range of technology implemenations and architectures, because network operators have diverse infrastructures and operational requirements. Mandating specific security controls might not suit every operator's unique environment or business model. Allowing flexibility enables operators to implement security measures that best fit their specific needs. Therefore, the 3GPP specifications aim to be broad and inclusive, accommodating various technological advancements and operator strategies.

Operational impact and cost considerations: Some security controls can introduce additional effort and complicate network operations; for example, for key management. While this is by no means a reason to simply forego security entirely, it may require operators to employ a risk-based approach for the implementation of security controls. To do so, operators need the freedom to balance security needs with budget constraints taking into account their individual risk profiles. Blankly mandating all security measures could be prohibitive for smaller operators or those operating in less affluent markets.

Lastly, it is important to recognize that international standards alone cannot effectively enforce security controls. As discussed at the beginning of this text, one of their primary purposes is to ensure global interoperability. This goal is achieved when the majority of operators agree on a mandatory set of measures required for effective interworking. Conversely, however, who would be responsible for mandating controls that are exclusively internal to a network operator's environment? These include measures such as access control for network operations and management or internal network segmentation. For these, there is simply no central authority to validate their implementation or effectiveness.

Therefore, while international standards play a crucial role in establishing a baseline for global interoperability, the responsibility for implementing and complementing them with additional measures as required falls on the individual network operators.

3.4 Conclusion

This chapter has discussed some of the optional 5G security controls and parameters in the 3GPP technical specifications. It highlighted various decisions available to both home network and serving network operators with regard to security as well as how those influence the protection of subscribers and their data. While some of them are arguably optimizations of existing functionality (e.g., SoR), others can be considered significant security improvements (e.g., SUPI encryption), if applied in practice. Both home network and serving network operators play an important role in determining the implementation of optional security controls and parameters. Not only do they have to decide on security controls within their individual network domains, some of these aspects also require collaboration between both parties to be effective.

Lastly, different factors were discussed for why certain aspects in international standards may be defined as optional and therefore left for implementors and network operators to decide. It can be concluded that there are indeed valid reasons for these limitations.

References

[1] 3GPP, TS 33.501, Security architecture and procedures for 5G System, 18.4.0, Jan. 2024.

[2] 3GPP, TS 33.210, Network Domain Security (NDS); IP network layer security, 17.1.0, Sep. 2022.

[3] 3GPP, TS 33.310, Network Domain Security (NDS); Authentication Framework, 18.2.0, Jan. 2024.

[4] 3GPP, TS 31.102, Characteristics of the Universal Subscriber Identity Module (USIM) application, 18.3.0, Jan. 2024

[5] 3GPP, TR 33.935, Study on detailed Long Term Key Update Process (LTKUP) detailed solutions, 17.0.0, Mar. 2022

Index

For Product Safety Concerns and Information please contact our EU representative GPSR@taylorandfrancis.com Taylor & Francis Verlag GmbH, Kaufingerstraße 24, 80331 München, Germany